Palgrave Studies in Literature, Science and Medicine

Palgrave Studies in Literature, Science and Medicine is an exciting, prize-winning series that focuses on one of the most vibrant and interdisciplinary areas in literary studies: the intersection of literature, science and medicine. Comprised of academic monographs, essay collections, and Palgrave Pivot books, the series will emphasize a historical approach to its subjects, in conjunction with a range of other theoretical approaches. The series will cover all aspects of this rich and varied field and is open to new and emerging topics as well as established ones.

Aaron Kitch

Shakespeare's Theater of Nature

Science, Religion, and the Orders of Mimesis in Early Modern Europe

palgrave
macmillan

Aaron Kitch
Bowdoin College
Brunswick, ME, USA

ISSN 2634-6435 ISSN 2634-6443 (electronic)
Palgrave Studies in Literature, Science and Medicine
ISBN 978-3-031-78081-3 ISBN 978-3-031-78082-0 (eBook)
https://doi.org/10.1007/978-3-031-78082-0

This Palgrave Macmillan imprint is published by the registered company Springer Nature Switzerland AG.
The registered company address is: Gewerbestrasse 11, 6330 Cham, Switzerland

For David Bevington, in memoriam

(His life was gentle and the elements
So mixed in him that nature might stand up
And say to all the world "This was a man.")

A Note on Texts and Spelling

For clarity and consistency, I have modernized spelling and punctuation of all early modern quotations, with the exception of some medieval titles and phrases that do not translate easily. All quotes from Shakespeare are from the revised Oxford text, edited by Stanley Wells and Gary Taylor, as printed in the Norton edition, 3rd ed., edited by Stephen Greenblatt et al. I cite this text parenthetically in the text by act, scene, and line numbers (just line numbers in the case of poetry). All quotations from Plato refer to the English translation in the *Collected Dialogues* ed. Edith Hamilton and Huntington Cairns (Princeton: Bollingen, 1989), cited parenthetically in the text by Stephanus numbers. All passages from Aristotle's works are derived from the translation of the Revised Oxford edition, ed. Jonathan Barnes, 2 vols. (Princeton: Bollingen, 1984), cited parenthetically by Bekker numbers, unless otherwise noted. All biblical passages refer to the Geneva translation based on the work of William Tyndale and first published in 1560—the edition that Shakespeare himself probably used—cited parenthetically in the text by chapter and verse.

Acknowledgements

I have accumulated many debts of gratitude in the research and writing of this book, which I dedicate to the memory of my graduate advisor, the late David Bevington. As anyone who worked with David knows, he was a model of good-natured thoughtfulness and gentle encouragement. My career as an early modern scholar would not be possible without the generous guidance of many additional teachers, including Margaret Ferguson, Richard Halpern, Carla Mazzio, Janel Mueller, Josh Scodel, and Richard Strier. For this project in particular, I especially appreciate the support of Mary Crane in the early stages of the book and William Cohen for saving me from some errors late in the production process. I am also grateful for the counsel and conversation of my co-editor of *Rethinking Science and Religion in Early Modern Culture* Jennifer Rust. I would also like to thank my editor at Palgrave, Molly Beck, for timely guidance and support, as well as the editors of the "Studies in Literature, Science, and Medicine" series, Sharon Ruston, Alice Jenkins, and Jessica Howell, for welcoming me to their list. I also thank the anonymous peer reviewers who provided invaluable feedback on both the book proposal and the completed manuscript.

At Bowdoin, I am fortunate to be surrounded by generous and enthusiastic colleagues, especially Aviva Briefel, Brock Clarke, Pete Coviello, Henry Laurence, Marilyn Reizbaum, Vin Shende, and Mark Wethli. I am also honored to have Ann Kibbie and Maggie Solberg as friends and fabulous fellow travelers in pre-1800 literature. Team-teaching classes on Shakespeare with Abbie Killeen in Theater and Dance and Paul Franco in Government and Legal Studies has been revelatory and rewarding. My colleague in Classics Mike Nerdhal has been a bountiful source of knowledge about all things

Latin and Greek. My thanks as well to those current and former students who have joined me on various early modern explorations, including an advanced seminar on "The Arts of Science in Early Modern England": Andrew Chang, Sarah Bonnano, Astrid Braun, Colleen Doucette, Paige Gill, Vincent Han, Lily Janas, Julia Jennings, Ella Jones, Francesca Kusserow, Mira Pickus, Dylan Richmond, Micaela Simeone, Ian Stuertz, Jeremy Tewari, Miranda Viederman, and Maddie Wolfert. Although Daniel Hennelly and Chris Zhang have not enrolled in my classes, they have been engaging interlocutors and inspiring teachers in their own right.

A Dibner fellowship in the history of science at the Huntington Library provided a stimulating year of research, conversation, and writing that proved foundational for this project. I am grateful to Steve Hindle and other members of the Huntington staff for their support during my stay in California, as well as to the other fellows that joined me there. A short-term fellowship from the Getty Library was also providential. I thank David Brafman in particular for sharing his knowledge and for helping me find useful books and objects. A Faculty Leave Supplement fellowship and several Fletcher Family research grants from Bowdoin were also vital for completion of the project. My research would not have been possible without the tireless assistance of librarians at the Huntington, Getty, Newberry, Folger, Houghton, and British Libraries. I thank them heartily. Closer to home, Bowdoin's library staff has also been extraordinary: thanks to Carmen Greenlee, Beth Hoppe, Jaime Jones, Kat Stefko, Marieke van der Steenhoven, and Guy Saldanha for their skill and support.

As always, my greatest thanks are to my family: to my mother, Sally, a trail-blazing scholar who has charted her own path through the academy; to my father, Tom, for his tireless encouragement and generosity; to my brother, Justin, and his family for the wine and music; to my wife, Allison, for her grace and good humor; to my children, Elsa and Beckett, for their wit and wisdom; and to my dog, Juno, who is a theater of nature unto herself.

An earlier version of Chap. 4 appeared as "Idolatry, Sexuality, and Protestant Theatricality in *The Two Gentlemen of Verona*," *Medieval and Renaissance Drama in England* 34 (2021): 81–101. © 2021 by the Rosemont Publishing & Printing Corp. An earlier version of Chap. 7 was published as "Shakespeare's 'Monstrous State': Divination, Typology, and the Book of Nature in *Julius Caesar*," *Modern Philology* 118.3 (2021): 340–63. © 2021 by the University of Chicago Press. I appreciate the permission from both presses to republish revised versions of this material here.

Contents

About the Author

Aaron Kitch is Associate Professor of English at Bowdoin College and the author of *Political Economy and the States of Literature in Early Modern England* (Ashgate / Routledge). Together with Jennifer Rust (St. Louis University), he is the co-editor of *Rethinking Science and Religion in Early Modern Culture* (Bloomsbury) and has published essays in *Studies in English Literature, Religion and Literature, Shakespeare Quarterly, Modern Philology,* and *Configurations,* among other journals.

List of Figures

CHAPTER 1

Introduction

Suit the action to the word, the word to the action; with this special observance, that you o'erstep not the modesty of nature: for anything so o'erdone is from the purpose of playing, whose end, both at the first and now, was and is, to hold, as 'twere, the mirror up to nature; to show Virtue her own feature, scorn her own image, and the very age and body of the time his form and pressure.
—William Shakespeare, Hamlet *(3.2.16–22)*

Hamlet's instructions to the players at Elsinore come as close as any lines by Shakespeare to outlining a defense of theater (*ars dramatica*). Hamlet's fascination with modes of representation matches his obsession with nature in this passage that praises plays for their capacity to hold a "mirror up to nature" and warns the players not to "o'erstep" the "modesty of nature" in their performance of the Mousetrap before the court. Hamlet also celebrates drama for its capacity "to show Virtue her own feature, scorn her own image, and the very age and body of the time his form and pressure." Such mirroring occurs in and through embodied performance, which itself depends on "nature":

> Oh, there be players that I have seen play, and heard others praised, and that highly, not to speak it profanely, that neither having th'accent of Christians

A. Kitch, *Shakespeare's Theater of Nature*, Palgrave Studies in Literature, Science and Medicine,
https://doi.org/10.1007/978-3-031-78082-0_1

> nor the gait of Christian, pagan, nor man, have so strutted and bellowed that I have thought some of nature's journeymen had made men, and not made them well, they imitated humanity so abominably. (3.2.26–32)

Actors for Hamlet embody the creative potential of nature itself. His pun on "abominably" (*ab-hominem,* inhuman), in fact, suggests that inadequate players reveal faults in the nature that formed them. Hamlet employs a Christian norm ("accent" and "gait") to measure such transgression, though elsewhere in the play he seems no more interested in religion than he is a fan of cosmetics or mothers who marry the brothers of their dead husbands. (Of course, as many scholars have noted, the play is actually quite invested in religious questions and ideas.[1])

After hearing one actor recite a tearful speech about the fall of Troy, Hamlet is amazed that a mere "player" could express such powerful emotion—that he could "force his soul so to his own conceit" in order to portray the fictional figure of Hecuba (2.2.472). Hamlet berates himself for his failures to act on his own and more immediate "motive" for tears (2.2.480). This representative capacity of theater exposes his failures to express his natural passion while also providing a way to test the story of the ghost about his father's murder. Rather than take up arms against a sea of troubles or assume the role of Senecan avenger against his father's death, Hamlet arranges for a modified dramatic performance before Claudius and his court at Elsinore. One reason for the Mousetrap, according to Hamlet, is theological—it will confirm whether the ghost who appeared to him in the form of his father was a devil or a true spirit (2.2.517–19). Another is empirical: the play will allow him to "observe" his uncle as he "rivet[s]" his eyes to his uncle's face (3.2.73, 78). Both motivations confirm that "the play's the thing" (2.2.523).

This book explores early modern nature as both a source and subject of scientific and religious knowledge as well as dramatic mimesis. It concludes that such nature was increasingly understood as a complex structure of representation itself in early modern Europe.[2] Like Shakespeare's works themselves, the nature he draws on is multitudinous and unruly, not yet subjugated to strict taxonomies and laws as it would be in later centuries. In order to begin to understand this kind of nature on its own terms, we must first remove the post-Romantic spectacles that have shaped contemporary definitions of the concept. Doing so requires us to leave behind the heady mixture of "Greek politics, French Cartesianism, and American parks" that defines modern nature, as Bruno Latour jokes.[3]

The ideological architectures of early modern nature are more rich and varied. They are also thoroughly enmeshed in theological concerns. For example, a nascent ecological consciousness in early modern England responding to deforestation and overfishing invoked both new forms of natural philosophy and a theological awareness of the divinely created order of nature or *ordo creationis*. Thus the mathematician, navigator, alchemist, astrologer, and physician John Dee made empirically precise calculations of the amount of fish lost to Dutch fishing fleets each year in his *Perfect Art of Navigation* (1577) while also claiming that overfishing is an affront to God who, in his "infinite goodness, sendeth us yearly evident tokens of the great abundance of this kind of good victuals, prepared purposely for this part of British Albion."[4] As these words suggest, Dee viewed nature as a divine gift that was also entwined with our everyday life.

As Jean Feerick notes, "the social, the cultural, and the human were still perceived to be *inside* nature" in pre-Enlightenment Europe, not yet "separated from it and abiding by a discrete set of principles."[5] It was only in the late seventeenth century that nature began to be considered as external to humanity. The natural philosopher Kenelm Digby, for example, developed a theory of the "oeconomy of nature" in the late seventeenth century that introduced a radical break between nature and humans, as Peter Remien has recently detailed.[6] Of course, "science" and "religion" are themselves vast and overdetermined categories. As Carla Mazzio notes, for instance, science before Newton included activities as diverse as rhetoric, mathematics, clothmaking, astrology, alchemy, and baking.[7] Like Christianity, moreover, which fractured into multiple and competing factions during the Reformation, early modern science encompassed an expansive yet contested constellation of beliefs and knowledge practices.

By "mimesis" I mean historically varied structures of representation that reframed nature. The term understands both form and content as mutually defining and relational. It also incorporates technological innovations such as the telescope and the printing press, as well as new forms of literary and dramatic representation. Mimesis in this sense captures the experience of witnessing words, bodies, and objects on the newly permanent stages of early modern London. I share with Eric Auerbach's famous study of the concept an interest in the philosophical underpinnings of such forms of imitation, but I challenge Auerbach's effort to define an objective and uniform "reality" that literary representation increasingly accesses over centuries.[8]

My claim in the pages below is that Shakespeare participated in a broad rethinking of both "nature" and the modes of representation that defined and supported it. Such rethinking involved theologians, secular authors, natural philosophers, medical doctors, and artists, among others. The "religion" of my title mostly refers to Christian texts, beliefs, and practices, though I also address other religious traditions on occasion. I argue that Shakespeare's centering of nature as a source of theatrical representation follows a period of intense cultural and epistemological upheaval, caused in part by the Protestant Reformation but also emerging from broader questions of theology and natural philosophy, including how forms of divinity become legible in the *ordo creationis.* I approach works of drama by Shakespeare and other authors as multifaceted mirrors of this historical process. For early modern playwrights, as we will see, the mimetic capacity of drama both draws on and extends the mimetic capacity of nature itself. The newly established outdoor amphitheaters like the Globe in early modern London joined technological innovations such as the telescope, microscope, and printing press that enabled new modes of representing nature. Together, they redefined nature itself as both a source and object of representation.[9]

Timothy Morton observes in *Ecology without Nature* (2009) that "nature always slips out of reach in the very act of grasping it."[10] This book acknowledges and historicizes his claim by arguing that nature was not just the subject of closer observation through mediating technologies such as the telescope and microscope or made new through words and images in print—it was also revealed as a structure of representation itself. Shakespeare both absorbed and transformed debates about nature as a source of scientific and religious knowledge in works of drama that are self-conscious about their status as modes of artificial representation. His plays mix art and nature in new ways by experimenting with relations between words, images, and objects as sources of knowledge and pleasure. These hermeneutic issues were also deeply embedded in questions of religious and political authority. In assessing the semiotics of such drama, for example, Robert Weimann has identified a "bewildering separation of visible signs from transcendental meanings" in in early modern England:

> Because authority, including the authorization of discourse itself, was no longer given, as it were, before the writing and reading began, the act of representation was turned into a site on which authority could be negotiated, disputed, or reconstituted. Modern authority, rather than preceding its

> inscription, rather than being given as a prescribed premise of utterances, became a product of writing, speaking, and reading, a result rather than primarily a constituent of representation.[11]

For Weimann, the Reformation forced a new reckoning with the word, rendering "discourse" a "site on which authority could be negotiated, disputed, or reconstituted." Shakespeare's theater for Weimann as for this study was one such site, but Weimann overlooks the significance of early modern scientific practices and the degree to which "nature" itself was part of this process of revising authority. Consumed by paying audiences seeking entertainment, Shakespeare's drama staged new relations between nature and God as versions of what Walter Benjamin calls "secularized ritual" that subjected doctrinal Christianity to contrasting models of mimesis.[12] Such drama exposes a conceptual tension between a more reflective and meditative relationship with the order of nature and a more active relationship that sought to make nature both visible and useful to humans in new ways. The Reformation motivated new uncertainties about established truths and the mediums that delivered those truths, producing what John Parker calls a "crisis of mimesis" that accelerated in the sixteenth century.[13]

The Nature of Mimesis

This book situates Shakespeare's theater of nature in relation to broader traditions of philosophical debate about the status of mimesis itself. Plato famously rejected forms of representation that remove us from what he considered to be the invisible forms of truth. In the *Republic* (c. 375 BCE), for example, Socrates compares mirrors that collapse three dimensions into two with paintings and sculptures that capture only external appearances rather than essence (596c-e). He also associates the human capacity to be entertained by "shadow-painting" and other forms of spectacle with political tyranny (602c-d). Rejecting Plato, Aristotle restored mimesis as a philosophical category, especially in his praise of tragedy in *The Poetics* (c. 330 BCE) as an imitation of an action that produces catharsis in sympathetic viewers (1450a). This foundational dialectic between Platonic formalism and Aristotelian mimesis shaped centuries of work in natural philosophy and theology. The introduction of new technologies of visual reproduction, including the printing press, reframed such debates in early modern Europe while also shifting relations between words, objects, and bodies as sources of representation. For example, Philip Sidney's *Defense of Poesy* (c. 1580) defines "poesy" as an "art of imitation" and a "speaking picture" with the aim "to teach and delight."[14] Sidney adopts philosophical ideas about both mimesis and nature from Aristotle and Plato in his influential account: the poet, whom Sidney refers to as the "monarch" of all human "sciences" (226), has a capacity to "grow in effect, another nature, in making things either better then nature bringeth forth, or quite a new form such as never were in Nature" (216). Such capacity depends for Sidney on the "works of nature":

> There is no art delivered to mankind that hath not the works of nature for his principal object, without which they could not exist and on which they so depend as they become actors and players as it were, of what nature will have set forth. So doth the astronomer look upon the stars, and by that he seeth, setteth down what order nature hath taken [...] the natural philosopher thereon hath his name, and the moral philosopher standeth upon the natural virtues, vices, and passions of man [...] The physician weigheth the nature of a man's body, and the nature of things helpful or hurtful unto it. And the metaphysic, though it be in the second and abstract notions, and therefore be counted supernatural: yet doth he indeed build upon the depth of nature. (215–16)

For Sidney, as for Shakespeare, "nature" is both the ground and horizon of imaginative writing. Poesy and natural philosophy are, in fact, mutually defining in his account.[15] Anticipating Shakespeare's language in *Hamlet*, poets, astronomers, physicians, and "metaphysic" philosophers for Sidney are "actors and players" of what nature provides (216). This definition of poesy is Platonic in the sense that poets have the freedom to transcend nature by creating fantastical figures and heroes—inhabitants of the "golden" world of art who are larger than life. Yet such poets also resemble Aristotelian natural philosophers in their need to go "hand in hand" with nature.

And they were not the only ones. As E.M. Gombrich observed some time ago, a number of early modern artists sought a "conquest of reality" by recovering and mastering classical techniques of visual illusion—including Albert Dürer and Georg Hoefnagel.[16] Their efforts to capture nature through modes of visual representation also celebrated the divine origins of the *ordo creationis*.[17] Like Shakespeare, early modern visual artists inherited numerous and sometimes conflicting ideas about the natural world from classical and medieval sources, including scholastic accounts that mixed Aristotelian natural philosophy with Christian theology. Ubiquitous tropes such as the "Book of Nature," the "Mirror of Nature," and the "Ladder of Nature" had figured the *ordo creationis* as a mode of textual representation for centuries, but they shifted their meaning in central ways in the period. Grounded in biblical passages such as Paul's observation in Romans that "the invisible things of God are clearly seen, being understood by things that are made" (1:20), the Book of Nature received its fullest treatment in Christian theology by Augustine, who implored readers of his commentary on the Psalms to "roam through the whole creation" in order to see how "everything cries out to you of its Author."[18] Augustine elevated the Book of Nature as a Neoplatonic structure that encouraged symptomatic reading of the natural world, a world valued not so much in itself as for how it represented the invisible world of divinity that it manifested. The medieval trope of the Ladder of Nature likewise employed a Platonic logic that shaped the concept of a divine chain of being leading from the tiniest, visible creatures to the heavens.[19] Over the course of the sixteenth and seventeenth centuries in England, however, these popular tropes gave way to a more complex vision of nature as divine matter that demanded new tools and modes of investigation in order to yield truths not just about God but also about nature itself.

A new trope also emerged in the sixteenth century: the "Theater of Nature." It was theorized most fully by the French jurist and philosopher Jean Bodin in his *Theater of Universal Nature* (*Universae naturae theatrum*, 1576). For Bodin, the theater of nature is "nothing other than a sort of table [*tabula*] of the things created by the immortal God placed before the eyes of everyone, so that we may contemplate and love the majesty, power, goodness, wisdom of the author himself and his admirable providence in the highest, the middling, and the smallest things."[20] Like the Book of Nature, the Theater of Nature regards the *ordo creationis* as a structure of divine revelation while also claiming that only natural philosophy gives a privileged view of such nature: "it alone can wrest, even from those who are unwilling, clear assent about the state and origin of the world and the infinite power of one eternal God."[21] Theatricality for Bodin is not physical or material but rather defines the representative capacity of nature itself that becomes visible and knowable in new ways through print.[22]

Bodin's work was not translated into English until the late seventeenth century, but it went through several Latin and French editions during

Shakespeare's lifetime. Bodin's work joined the *Theatrum mundi* of Pierre Boaistuau that was first published in 1566 and went through at least four editions before 1620. Like Bodin, Boaistuau was a French polymath who wrote in multiple genres and approached nature as a source of divine creation, a quality of human behavior, and a means of reproducing human life. Boaistuau presents his treatise as a Christianized form of Pliny's natural history that serves several explicit moral purposes. The first was to elevate humankind from a natural state of corruption and sin by teaching us to know ourselves as "a similitude or statute in this world, which is a very shop of the works of God" (sigs. B5v-B6). The second was to feel humility in the face of divine creation, as readers recognized the superior capacity for self-sufficiency of animals and their ability to live within the bounds of nature (B6–B7v). Both claims resonate with Shakespeare's drama on several levels. Indeed, Boaistuau's translation of "Rhomeo and Juliet" from the Italian was likely Shakespeare's primary source in writing his dramatic take on the classic tale.[23] Like Bodin, moreover, Boaistuau's *Theatrum mundi* self-consciously situates art, nature, and divinity in relation to one another. For example, Boaistuau argues that there "is no Art nor science but that men have excelled each one in their degree, more or less according to the influences and favors that hath been given them from heaven" (sig. K6) and praises magnificent inventions, including a giant mirror or "glass" made by the Persian King Sabot on which he could sit and view the heavens spread out in inverted form (sig. R8v). But the greatest of all such "miracles of nature" for Boaistuau is the printing press, which "alone hath entered with such good hap and perfection into this world, that there cannot be added nor diminished anything that doth not render it defective or deformed." The press not only immortalizes human acts of writing, it "eterniseth world without end" (sig. S2).

Boaistuau's view of the "art of printing" as a divine discovery in nature was widely shared in the sixteenth century, including by Bodin.[24] Printing celebrated the order of nature as divine creation and made it visible in new ways. Print technology also helped to produce what Paula Findlen calls the "expansion of nature's metamorphic potential" in the period.[25] As Findlen argues, such potential depended on a "reflexive system of signs, symbols, and correspondences which created horizontal relations among kingdoms and worlds."[26] Before natural "laws" became the primary object of natural philosophy in the seventeenth century, "nature" was an expansive, unruly, and untamable presence that was both available for study and inevitably obscure and unpredictable. Some early modern natural philosophers embraced Plato's definition of nature in the *Timaeus* (c. 360 BCE) as the product of a distant and inscrutable demiurge, while others preferred Aristotle's focus on the material composition of the visible world (even though nature for Aristotle is both material and immaterial). But nature for early modern authors as for Aristotle was incomplete and imperfect—it required human intervention. In the *Politics* (c. 350 BCE), for example, Aristotle notes that "art and education" both imitate and complete what nature began, compensating for "Nature's deficiencies" in the process (1337a).[27] Aristotelian nature inspires human activity without circumscribing such agency. It also combines both human and non-human entities. Objects such as stones thrown into the air, for example, "seek" to return to the ground because they desire to return "home."[28] It is not surprising, then, that non-human forms of nature became a subject of inquiry and literary treatment for many medieval and early modern authors.[29]

Contrary to later claims about the historical and inevitable conflict between science and religion, moreover, natural philosophy and Christian theology were mutually defining in early modern Europe. The challenge of Protestantism to Rome as the ultimate authority in matters of religious belief and ritual, for example, prodded new modes of inquiry into the natural world, which in turn helped to generate new forms of natural theology. Reformers sought to strip away allegorical accretions in efforts to access the direct Word of God, in both the bible and the natural world. They also highlighted questions about the relative status of images, objects, and words as mediums of divine power.[30] It is perhaps no coincidence that the chief architect of scientific knowledge in England, Francis Bacon, was raised by a strong Protestant mother, educated by a Puritan tutor, and attended Cambridge at a time when it was a hotbed of Calvinism. Bacon pursued science as a project of reform that could subjugate nature in new ways in order to understand it, liberating mankind from the curse of the Fall in the process. Bacon longed to chain nature in order to make it submit to human needs. In his *Advancement of Learning* (*De Augmentis Scientiarium*, 1605), for example, Bacon describes "free" (*libera*) nature in its ordinary course as the source of the order of nature itself before invoking the myth of Proteus to argue that natural philosophers should bind and vex nature in order to force it into new forms. By using "artificial things" such as microscopes, "nature is held in subjugation by the empire of man, for without man these things would never have been made." These technological innovations allow us, Bacon claims, to see bodies in an entirely new guise as "a kind of alternative universe or theater of things."[31]

The aims of science and art also overlap for Bacon: to bind and vex nature through natural magic and experimentation. Art joins with nature, for example, to produce new hybrids by creating artificial representations of natural universes. Technological innovations such as the microscope and the telescope could both correct for the natural deficiencies that troubled Bacon and uncover structures of representation *within* nature itself. The fullest account in England of these visual innovations was provided by Robert Hooke in his *Micrographia* (1665), but experiments with enhanced visual technology began in the previous century, when for example Giordano Bruno remarked that "he who does not imagine in words and pictures is no philosopher" (*non est philosophus, nisi qui fingit et pingit*) and when illustration was newly central to modes of scientific inquiry.[32] Hooke's catalogue of images and "descriptions of minute bodies made by magnifying glasses" regards the microscope as more than just a technological wonder. Magnification of the order of nature allows him to "discern all the secret workings of Nature, almost in the same manner as we do those that are the productions of Art, and are managed by Wheels, and Engines, and Springs, that were devised by human wit."[33] Looking closely at nature through the microscope, Hooke invites his readers to consider the ways in which the order of nature resembled mechanical "productions of Art."[34]

One paradox of these efforts to return to nature in early modern Europe was that the more closely natural philosophers tried to observe nature in and of itself, the more they discovered its reliance on forms of representation. Nature, it turned out, was always already embedded in language, a truth revealed even more fully with the scientific discovery of DNA as the genetic code of all life on the planet. [35] It was a pipe dream of many seventeenth-century natural philosophers following Bacon that words (*verba*) might disappear directly into the things (*res*) that they signified, eliminating the signal noise of language as a medium.[36] Such a utopian desire motivated the schoolteacher John Webster, for example, to claim in the 1650s that nature was superior to books as a source of knowledge because it promised to stabilize the semiotic uncertainties of written and oral language. Based on this assumption, Webster urged Oxford and Cambridge to teach the languages of nature rather than "knowledge of tongues" because such natural "*Hieroglyphicks*" could convey the "invisible power" and "divine wisdom" of God in a way that was more directly beneficial to humankind than the study of human languages.[37] Although Webster's recommendations were never adopted, his ideas resonated with members of the Royal Society who believed that their work could uncover the laws of nature to reveal the "divine wisdom" of God as the creator while moving away from the dangers of figurative language. The motto of the Royal Society, founded in the name of Bacon in 1660, was *nullius in verba*, a phrase invented by John Evelyn as an adaptation of Horace that he invoked to reject the bondage of the past. The phrase roughly means "not bound by words," though "words" in this sense refer primarily to authorities such as Aristotle and scholastic commentaries on him.[38] In lieu of such textual authorities, Evelyn and the other founding members of the Royal Society hoped to produce nature in and of itself—nature unbound.

Shakespeare's drama and the institution of professional theater for which he wrote emerged from the same cultural milieu that motivated Bacon and the members of the Royal Society, but where Baconian natural philosophers sought to bury words in things in a failed effort to gain unmediated access to nature, Shakespeare and his fellow playwrights embraced linguistic wordplay and doublespeak as sources of poetic and dramatic power. Far from seeking to substitute words for objects, Shakespeare drew out the creative power of metaphor and figurative language to forge self-reflective characters whose identities depended almost entirely on words. Shakespeare's audience reveled in verbal friction and bawdy wordplay, just as they wondered at the imaginative worlds sustained by such wordplay and enacted by what Hamlet calls "Nature's journeymen." At the same time, as the chapters below delineate, Shakespeare's theater of nature engaged some of the same philosophical and theological questions that motivated Bacon and others to reframe nature in new ways. For example, Shakespeare alludes directly to the trope of the Book of Nature in the first act of *Antony and Cleopatra* when a soothsayer recommended by Charmian to Cleopatra brags: "In nature's infinite book of secrecy / A little I can read" (1.2.7–8). This reference aligns fortunetelling with modes of gossip and ignorant forecasting in Cleopatra's court, but elsewhere in his drama Shakespeare addresses the questions behind the trope more fully and seriously, as my examination of augury and divination in Chap. 6 demonstrates below. Charmian's use of the Book of Nature invites us to consider Shakespeare's plays in relation to the enduring power of figurative language as a tool of knowledge about the order of nature.

Methods and Models

This book both builds on and challenges previous scholarship, including works of ecocriticism.[39] I join Robert Watson, for example, whose *Back to Nature: The Green and the Real in the Late Renaissance* (2017) situates poetry, drama, and painting in early modern Europe in relation to questions of empiricism, epistemology, theology, and semiotics. Like Watson, I argue that theories of language and theories of religion were interrelated and that debates over idolatry and transubstantiation are also debates about how to know the presence of the divine in the order of nature. Yet, unlike Watson and others, I do not identify "nostalgia for unmediated contact with the world of nature" as a central obsession of the period.[40] I also reveal early modern nature as more contested and layered than he acknowledges.[41] For example, I go beyond Watson's focus on "green" forms of landscape and his claim that pastoral as a genre is defined primarily by nostalgia for unmediated contact with the "real." I likewise challenge the equivalence of "nature" with agrarian landscapes that Charlotte Scott proposes in *Shakespeare's Nature* (2014).[42] My account, by contrast, expands nature as a concept to understand the theological, scientific, and ideological layers of early modern nature. Such an approach extends ecology beyond the usual suspects of the forest of Arden in *As You Like It* (c. 1599), joining ecocritics such as Jason W. Moore who describe the "web of life" in which nature is "inside us" and "around us."[43] This approach also overlaps with Steven Swarbrick, whose account of *The Environmental Unconscious* (2023) proposes a materialism without matter and employs Lacanian psychoanalysis to demonstrate the ways in which the natural world was structured like a language.[44] In detailed and engaging readings of poetry by Spenser, Ralegh, Marvell, and Milton, Swarbrick challenges the assumption of modern ecocriticism that "matter and meaning must coincide" by paying attention to the "disarticulating force of matter—its semiotic excess."[45] Like Swarbrick, my account incorporates both human and inhuman worlds, following new materialist and posthumanist criticism that deconstructs such binaries.[46] I define nature as both material and immaterial, as a play of forces that sometimes eludes human capacity to know and understand it.

Shakespeare's plays both draw on and reflect the capacity of nature to be a mirror of itself. In this claim, I question Harry Levin's argument that early modern mirrors were always "ethical" and conceptual rather than "realistic," though I agree with Levin that nature is both ideological and material.[47] I also revise Roland Mushat Frye's claim that Shakespeare's works were completely inward facing manifestations of "art as autonomous" and that they were "independent of theological systems."[48] In doing so, I integrate Shakespeare's drama into historically produced and ecologically attuned epistemes, sharing with Todd Borlik a sense that much ecocritical scholarship remains trapped between the Scylla and

Charybdis of "neo-Ovidian primitivism and neo-Hobbesian barbarism."[49] The first pictures early modern natural environments as Edenic spaces of purity while the second casts early moderns as ignorant abusers of the natural world—especially those Baconian natural philosophers who plunder the earth in the name of science, as Carolyn Merchant has argued.[50]

Nature was neither simply material reality writ large nor a homogenous ideological construction but, rather, a complex semiotic system whose legibility was the subject of renewed debate in the period in which Shakespeare lived.[51] I show how nature was both subject to new forms of observation and embedded in systems of representation whose logic remained shaped by Christianity. As Lorraine Daston notes, nature in early modern Europe combined the "art of God" with a self-referential "artificial imitation of itself."[52] By foregrounding critical intersections between early modern religion and natural philosophy, moreover, I also address critical lacunae in some ecocriticism that ignores vital theological frameworks of early modern nature. Such an approach also bridges two recent "turns" in early modern scholarship that have mostly occurred independently of one another: the turn to science and the turn to religion. In turning to science, numerous studies in recent decades have rejected the binary opposition between the "two cultures" of science and literature as described by C.P. Snow in the 1950s in order to accept Carla Mazzio's invitation to address "the distinctly narrative, literary, and hypothetical dimensions of experimental science and, conversely, the scientific dimensions of the literary or fictional experiment."[53] But many studies of literature and science in early modern England continue to downplay or ignore completely the religious ideas and debates that were historically central to science. As Mary Thomas Crane notes in her study of literature and the "new science" in early modern England, skepticism toward Aristotle's account of the natural world led to new scientific models of the world, but these models in turn caused growing division between older and more intuitive models of nature and the newer, more counterintuitive models that violated common sense and increased anxiety about the legibility of nature.[54] Although Crane acknowledges religious contexts in several places in her valuable study, she and many other scholars of science and literature bracket religious questions and contexts.

As a contribution to the study of the history of science, moreover, this book emphasizes the centrality of representation in scientific thought and practices in ways that historians of science mostly associate with later periods.[55] Such an approach aligns with recent efforts by historians of science to rethink the role of religion.[56] The celebrated historian of early modern alchemy Lawrence Principe, for example, has noted that "theological motivations," including a "desire to read [...] the Book of Nature," were the "single greatest driving force for scientific inquiry throughout the entire early modern period."[57] Yet the belief that religion and science were intractable enemies—drawing especially on the persecution of Galileo by the Inquisition—still endures, shaping popular accounts of religion as an obstacle to rather than a catalyst for scientific inquiry.[58] Like historians of science, scholars have also addressed secular assumptions in literary studies. New historicism, which drew heavily on cultural anthropology and the work of Michel Foucault, often bracketed religious questions or reduced religion to ideology. Modern historians of science take early modern religion more seriously, identifying its significance even in authors such as Shakespeare who appear to be purely secular.[59]

The revival of interest in religion has also been a vital force in early modern studies. Deborah K. Shuger, for example, understands religion as "the primary language of analysis" in early modern England.[60] Huston Diehl and Jeffrey Knapp have also studied early modern drama as a form of Protestant hermeneutics and community building.[61] A number of scholars have likewise read Shakespeare's plays as forms of political theology.[62] Even scholars who insist on the essentially secular function of early modern drama acknowledge that theological questions and modes of representation remain foundational.[63] At the same time, this book avoids speculating about Shakespeare's personal religious views.[64]

As my title indicates, this project situates Shakespeare in relation to "early modern Europe" rather than just England. In doing so, I join Noémie Ndiaye in adopting a more transnational approach to early modern studies, acknowledging in the process the problematic category of "Europe" as a geographical term that has never achieved cultural, religious, or political uniformity.[65] Featuring texts by authors produced both within and beyond England, I also suggest that some influential contexts for understanding Shakespeare's plays originate far beyond his place and time. Thus I sometimes analyze texts and debates from classical and medieval Europe that have been underappreciated in previous studies. At the same time, I recognize both continuities and discontinuities between classical, medieval, and early modern authors and intellectual traditions.

My interest in the history of print technology joins continuing debates among scholars from many different backgrounds about the impact of the printing press on European culture, politics, and science. Elizabeth

Eisenstein's influential account of *The Printing Revolution in Early Modern Europe* (1979) has received fierce criticism from Adrian Johns and others for its sweeping claims about printing, especially in relation to early modern science, yet some of her more prescient claims have been lost in the backlash, including her arguments about print's relation to the study of nature.[66] As Eisenstein argues in the final chapter of her study, the proliferation of printed data and models of nature encouraged natural philosophers such as Tycho Brahe to be more precise in their calculations and to share calculations more easily. Her claim is not just that printing "fixes" information completely and permanently but that it also reframed the order of nature.[67] More recent work by historians of science has echoed and reinforced these findings. Sachiko Kusukawa, for exmaple, notes that print was the "primary medium" through which many natural philosophers "made known their discoveries and arguments."[68] And Paula Findlen highlights the capacity of the press "to circulate words and images more widely," helping to enhance both the status and reach of natural philosophy in the period.[69] Engraved and woodblock images incorporated into printed books and pamphlets also claimed a new epistemological authority as what the anatomist Andreas Vesalius called a new mode of "understanding" in the sixteenth century.[70] As I demonstrate in the following chapters, the visual technology of print in early modern Europe shifted relations between verbal, visual, and spoken signs.[71]

My approach to drama as a visual and embodied mode of performance likewise draws on recent work by historians of science on the history of vision.[72] In *The Body of the Artisan* as well as in her contribution to an anthology of essays on the "Science and Visual Culture," for example, Pamela Smith has emphasized the role of image makers in advancing knowledge of the natural world.[73] Likewise, Pamela O. Long explains how "visual representation came to legitimate knowledge claims" in early modern Europe, noting that the "artifactual and the natural moved into closer proximity" as a result.[74] Such an approach to visual culture resonates with W.J.T. Mitchell's notion of the "the social construction of visual experience." Although Mitchell largely avoids religious questions in his influential work on visuality, his definition of "iconology" as "the study of images across various media" demands attention to technological innovation, social practices, and forms of representation.[75] My approach diverges from those scholars who define Shakespeare's visual drama in terms of emblem books, though it does not necessarily contradict this approach.[76] Still, I propose a different understanding of visuality as shaped by early modern natural philosophy and debates in theology. At the same time, I try to avoid an overly deterministic approach to the study of technology as found, for example, in works by Walter Ong and Marshall McLuhan.[77] I follow instead the Marxist model of Walter Benjamin, who understands changes in form, content, and distribution of film and literature in relation to technological change. In this regard, it is important to remember that, even as Benjamin focuses primarily on the historical emergence of photography and film, he also addresses earlier traditions of drama and printing.[78]

Chapter Summaries

The following two chapters provide a foundation for the rest of the book by addressing important intersections between religious, scientific, and mimetic traditions before Shakespeare. Chapter 2, "Representing Nature," studies new representations of nature in print as part of a larger hermeneutic revolution in early modern Europe. Both classical and early modern works of natural philosophy joined this process, including Pliny's famous encyclopedia of natural history that remained a central source of knowledge about nature for both formal and everyday scientists over many centuries. I show how influential "books of nature" printed in the sixteenth century reconfigured relations between words and images as modes of knowledge about the natural world, shaping Shakespeare's representation of Friar Laurence in *Romeo and Juliet* (c. 1594), for example.[79] I also highlight questions about how to understand and represent God that shaped such print production. Chapter 3, "Staging Nature," examines earlier traditions of theater in England that foregrounded problems of theatrical representation, including liturgical tropes, cycle plays, morality plays, and interludes of the early sixteenth century. I focus in particular on the innovative courtly interludes of Henry Medwall, John Rastell, and John Heywood, among others, that foregrounded nature in new ways in early Tudor England. The chapter ends by considering how Marlowe's *Doctor Faustus* (c. 1589) and Shakespeare's *Romeo and Juliet* reconfigure these questions of nature while also developing new techniques of theatrical representation.

Subsequent chapters turn more fully to Shakespearean drama, beginning with Chap. 4, "The Nature of Idolatry in *Two Gentlemen of Verona*," which approaches idolatry as a complex and overlapping problem of theology, natural philosophy, and mimesis. Influential Protestants such as Martin Luther and Philip Melanchthon drew on new scientific practices in their attacks on idolatry in sixteenth-century Europe, while Francis Bacon identified idolatry as a multi-faceted obstacle to scientific knowledge. Idolatry also had a shaping influence on theatrical practices from medieval manuscripts to the emergence of the new, permanent theaters of London in the sixteenth century. Shakespeare focuses on idolatry as a problem of theatrical representation in the famous reanimation scene of Hermione's statue at the conclusion of *The Winter's Tale* (c. 1611), but he explores idolatry more fully as a source of dramatic mimesis in his early comedy of *Two Gentlemen of Verona* (c. 1593), where he uses it to define the central problem in the play of how to represent love. I argue that Shakespeare's early comedy resonates with scientific and religious debates about the real "presence" of bodies in nature and of Christ in the eucharist, helping to redefine the status of words, images, and objects as tools of theatrical representation.

Chapter 5, "Natural Perspectives in the Henriad," analyzes the importance of visuality to questions of human and divine ontology, politics, and theatrical representation. In contrast to the arts of perspective as theorized by Leon Battista Alberti and others, Shakespeare's drama is visually diffracted and decentralized. History plays such as *Richard II* (c. 1595) and *Henry V* (c. 1598) help to construct what Francis Bacon calls a "perspectival" vision of history by foregrounding conditional and artificially constructed stage action. Representing forms of nature in *Richard II* also produces contingent and partial vision that disrupts overarching metanarratives of divine history. In *Henry V*, the duality of Henry as a figure of both Machiavellian calculation and heroic transparency emerges in the play by means of a bifurcated, shifting, and relational gaze of the audience that must interpret his character both in terms of what is and is not visible in performance.

Chapter 6, "Natural Medicine in *All's Well that Ends Well* and *Macbeth*," highlights medical texts and debates as they contributed to new approaches to the divine and mimetic structures of early modern nature. The anti-Galenic theories of Paracelsus (born Philippus Aureolus Theophrastus Bombastus von Hohenheim) were slowly gaining acceptance at courts around Europe, but the physicians of the Royal College of Medicine still generally looked down on anyone who worked directly with bodies, including surgeons and mostly female "empirics."[80] These facts are well known to historians of medicine, but many scholars have not appreciated how such debates about modes of healing navigated different understandings of language and the divine order of nature. Shakespeare reflects elements of these debates in scenes of healing in *All's Well that Ends Well* (c. 1605) and *Macbeth* (c. 1606). The cure of the king that occupies most of the first two acts of *All's Well that Ends Well*, for example, recalls Paracelsus, who integrated the study of the order of nature into his medicine. It is also part of a larger concern in the play with relations between visible and invisible phenomena. In a similar vein, the words and actions of Lady Macbeth can only be understood by an unseen and unrepresentable "Heaven," leaving Shakespeare's theater to register the visual effects of an unknown divine force in an uncertain and limited fashion. Simon Forman's intense scrutiny of this tragedy in his eyewitness report on Shakespeare's plays responds to the play's fascination and ultimate frustration reading the signs of the medical body.

Like medicine, early modern divination in its many forms—including astrology, augury, and chiromancy—encouraged close observation of the *ordo creationis*. Chapter 7, "The Nature of Divination in *Julius Caesar* and *Antony and Cleopatra*," identifies the overlapping functions of mimesis, religion, and representations of nature in scenes of divination in *Julius Caesar* (c. 1599) and *Antony and Cleopatra* (c. 1606). The historical figure of Julius Caesar was central to classical, medieval, and early modern narratives of divination, especially in relation to popular stories about a comet that was said to burn for nine days and nights following his death. Shakespeare alludes to the comet in *Hamlet* (c. 1600) and *Julius Caesar* without staging it directly. But his account of Caesar frames divination in terms of limitations of knowledge associated with efforts to understand the order of nature. The chronological confusions of divination also complement the typological structure of *Julius Caesar* and *Antony and Cleopatra* in which the death of Julius Caesar gives way to the universal peace under Augustus that makes the birth of Christ possible.

Chapter 8, "*King Lear* and the Orders of Nature," situates Shakespeare's bleakest tragedy in relation to the central ideas of this study, challenging a critical binary that reads the play as either spiritually redemptive or bleakly materialist. Focusing on Shakespeare's elevation of the order of nature allows us to see how material and spiritual meanings intertwine in the play, which was first performed around 1605. I argue that Lear's journey through the barren landscape of the second half of the play registers an epistemological shift from a transcendent system of pagan theology in which language, nature, and the divine are unified in Lear's voice to an immanent and materialist order that gives rise to new forms of theatrical performance and that generates the spiritual reformation of central characters. Lear's trajectory in the play from divinely supported monarch to "bare, forked animal" (Q 3.4.95 / F 3.4.100) depends on his passage through the order of nature, his use of modes of natural philosophy, and different genres of theatrical mimesis. Beginning with Lear's invocation of a pre-Socratic idea that "nothing will come from nothing" and moving through questions of astrology, meteorology, anatomy, and natural medicine, modes of natural philosophy provoke key shifts in both theological questions and the spiritual states of characters. Shakespeare's expansive tragedy unites many strands of argument that the book has been pursuing, illustrating foundational links between science, religion, and the orders of representation.

A brief epilogue expands on several central questions of the project, including the imaginative "fecundity" of early modern nature as framed by Daston's *Against Nature* and the implications for my study in relation to Charles Taylor's account of the history of secularism. I also address Shakespeare's mode of "indirect theology" in relation to some of his contemporaries, including Christopher Marlowe and Ben Jonson, who adopted a much more direct and polemical stance toward religious subjects in their plays.

Notes

1. *Hamlet* raises questions of sin, providence, and the afterlife, among other topics. See, e.g., Alan Sinfield, "Hamlet's Special Providence"; Philip Edwards, "Tragic Balance in *Hamlet*"; Catherine Belsey, *Shakespeare and the Loss of Eden*, 129–74; Heather Hirschfeld, "Hamlet's 'First Corse'"; Phoebe S. Spinrad, "The Fall of the Sparrow and the Map of Hamlet's Mind"; and John Gillies, "The Question of Original Sin in *Hamlet*." Robert Watson suggests that *Hamlet* "deeply condemns the illusions of afterlife it superficially encourages" in *The Rest Is Silence*, 74. Peter Lake observes that Hamlet "talks to himself" in his soliloquies "like an atheist," but also situates the play more broadly in relation to post-Reformation anxiety over providence and early modern conversion narratives in *Hamlet's Choice*, 135.
2. Desiree Hellegers notes that debates about nature were also "over biblical interpretation and over the nature of language itself" in *Handmaid to Divinity*, 10.
3. Latour, *The Politics of Nature*, 5.
4. See the excerpt by Dee on "Manifold disorder used about fry and spawn" from *The Perfect Art of Navigation*, reprinted in Borlik, ed., 354.
5. Feerick, "Economies of Nature in Shakespeare," 35–6.

6. Remien, *The Concept of Nature in Early Modern English Literature.*
7. Mazzio, "Shakespeare and Science," 1–2.
8. Auerbach mainly uses "reality" to refer to the everyday experience of everyday people. He claims that such everyday experience only becomes fully legible in certain novels of the nineteenth century. For an analysis and expansion of this principle, see Frederic Jameson's discussion in *The Antinomies of Realism*, 141–3.
9. For one version of this redefinition, see Catherine Wilson, *The Invisible World.*
10. Morton, *Ecology Without Nature*, 19.
11. Weimann, "Representation and Performance," 500.
12. Benjamin is speaking, of course, of the loss of the aura in the age of photography, film, and other modes of "mechanical reproduction." His claim, however, that the "ritual function" of art "is still recognizable as secularized ritual even in the most profane forms of the cult of beauty" is also relevant in earlier eras, especially with regard to the mechanical reproduction of the printing press, which Benjamin also mentions. See "The Work of Art in the Age of Mechanical Reproduction," e.g., 224, 231–2.
13. Parker, *The Aesthetics of Antichrist from Christian Drama to Christopher Marlowe*, 61.
14. Philip Sidney, *The Defense of Poesy* (c. 1579), in *Major Works*, ed. Duncan-Jones, 216–17. Hereafter cited parenthetically in the text by page number.
15. Technically, "natural philosophy" refers to those branches of "natural science" that derive from Aristotle's *libri naturales* ("natural books"), specifically those fields of inquiry associated with the movement of bodies and with change over time. These included mathematics and fields of study such as optics and astronomy that were associated with mathematics, but since both optics and astronomy could also include the study of moving bodies, the distinction was not clear cut. For more, see Domingo Gundisalvo, *On the Division of Philosophy*, trans. Clagett and Grant, 62–5. See also David C. Lindberg, "Science and the Early Church," 70. n. 1; and the "Introduction" to *Natural Philosophy Epitomised*, trans. and ed. Cunningham and Kusukawa, xlvi–xlix.
16. Gombrich applies the phrase "conquest of reality" to fifteenth-century art in *The Story of Art*, esp. 223–46.
17. Expanding on Gombrich, Thomas DaCosta Kauffmann captures this divine sense in his essay on the "Sanctification of Nature" in *The Mastery of Nature*, 11–48.
18. Augustine, *Augustine on the Psalms*, 1: 272.
19. Examples of the *scala natura* can be found in Boethius's *De consolatione philosophiae* (523 CE) and in the writings of St. Bonaventure. For an overview, see Kellie Robertson, *Nature Speaks*, 55–7. Arthur Lovejoy also

examines the trope in his *The Great Chain of Being*, 58–9. For rebuttals of Lovejoy, see William F. Bynum, "The Great Chain of Being After Forty Years: An Appraisal," and John North, "Some Weak Links in the Great Chain of Being."

20. Jean Bodin, *Universae naturae theatrum*, sig. 3v. Trans. Ann Blair in *Theater of Nature*, 160. As Blair notes, "in the theater of nature the human spectator does not become an actor with a moral dimension, or the object of the spectator's passion [....] instead, in the *Theatrum* he praises the contemplation of the constancy, beauty, and harmony of unadulterated nature as the most direct route to the divine Creator" (158–9).
21. Bodin, *Universae naturae theatrum*, in Blair, 22.
22. For Blair, Bodin's theater metaphor "emphasize[s] a visual, nonverbal form of contemplation" (154). She also notes that books of natural philosophy and theaters of anatomy were understood as "theaters" of knowledge in the period (155–61), but she overlooks the importance of the emergence of public playhouses in Europe and downplays the important role of images in similar works. For a different account of the *theatrum mundi* in relation to theatrical practice and forms of natural philosophy, see William West, *Theatres and Encyclopedias in Early Modern Europe*.
23. As René Weis suggests in his Arden 3 edition of *Romeo and Juliet*, Shakespeare most probably read the story of "Rhomeo and Julietta" in William Painter's *Palace of Pleasure* (1567), which was also a source for several other plays, but his central source for the tragedy was likely the *Tragical History of Romeus and Juliet* translated by Brookes in 1562 and based on Boaistuau's translation of Matteo Bandello. See *Romeo and Juliet*, ed. Weis, 44.
24. For Henry Bullinger, for example, the art of printing is "a most excellent & profitable gift of God given unto us in these last days." See *An Exhortation to the Ministers of God's Word*, sig. J2. Andreas Hyperius describes printing as "found out by the providence of God" in *The Course of Christianity*, 66. In *The Cosmographical Glass* by the "doctor in physic" William Cuningham, the "discovery" of print is attributed to John Faustus in Germany (sig. R1v).
25. Paula Findlen, "Jokes of Nature and Jokes of Knowledge," 316.
26. Ibid., 325.
27. For an overview of nature in Aristotle, see Arthur O. Lovejoy, "Nature as an Aesthetic Norm," 69–77, and the section on "'Nature' as Norm" in the *Documentary History of Primitivism and Related Ideas*, ed. Lovejoy et al. 1:185–91.
28. Robertson, *Nature Speaks*, 8.
29. The classic study of medieval representations of nature is George Economou, *The Goddess Natura in Medieval Literature*.

30. For accounts of such debates, see *Sacred and Secular Transactions in the Age of Shakespeare*, ed. Brokaw and Zysk, and Sara Ritchey, *Holy Matter*. Peter Harrison notes that the Book of Nature was, in part, an "attempt to stabilize the realm of the visible." See "The Book of Nature and Early Modern Science," 4.
31. Bacon, *The Advancement of Learning*, ed. Kiernan, 321. See also Sophie Weeks, "Francis Bacon and the Art-Nature Distinction." Bacon, *The Great Instauration*, ed. Rees and Wakely, 455. See also Weeks, 129.
32. Bruno's remark was made in a treatise published in London in 1583, cited in Christoph Lüthy, "What Does a Diagram Prove that Other Images Do Not?," 227. Martin Kemp also remarks that early modern Europe witnessed the "rise of illustration as a major tool of science" in "Seeing and Picturing," 363.
33. Hooke, *Micrographia*, sig. a2v. See also William Eamon's discussion of this passage in *Science and the Secrets of Nature*, 296.
34. Hooke, *Micrographia*, sig. A3v (fol. 8).
35. The director of the Human Genome project, Francis Collins, refers to DNA as the "instruction book" of life, for example, in *The Language of God*, 52.
36. As Pavneet Aulakh argues, Bacon embraced energetic language or *enargeia* as a way of producing what he calls "lively images," especially in fables and other forms of literature that allowed him to enhance, in Aulakh's words, "the sensuous and image-making potential" of language itself. See "Seeing Things Through 'Images Sensible,'" 1154.
37. Webster, *Academiarum examen, or, The Examination of Academies,* 18–19.
38. For an account of the motto of the Royal Society in the context of the drive to eliminate figurative language by its founders, see Clive Sutton, "'Nullius in Verba' and 'Nihil in Verbis,'" 55–64.
39. For an overview of the evolution of ecocriticism, see Cheryll Glotfelty's "Introduction" to *The Ecocriticism Reader*, xviii.
40. Watson, *Back to Nature*, 5.
41. Efforts to expand the scope of ecocriticism include the essays in *Beyond Nature Writing*, ed. Kathleen R. Wallace.
42. Scott, *Shakespeare's Nature: From Cultivation to Culture.*
43. Moore, *Capitalism in the Web of Life*, 2–3.
44. Swarbrick, *The Environmental Unconscious.*
45. Ibid., 2.
46. See Karen Raber's overview of these ideas in *Shakespeare and Posthumanist Theory.*
47. Levin, *The Question of Hamlet*, 157.
48. Frye, *Shakespeare and Christian Doctrine*, 268.

49. See Borlik's Introduction to *Literature and Nature in the English Renaissance*, 4.
50. Carolyn Merchant, *The Death of Nature*. Merchant's claims about early modern science have been challenged more recently by Margaret Osler in "The Gender of Nature and the Nature of Gender in Early Modern Natural Philosophy."
51. Watson addresses the influence of the back-to-nature movement on Shakespeare's *As You Like It* in *Back to Nature*, 77–107.
52. Lorraine Daston, "The Nature of Nature in Early Modern Europe," 166.
53. C.P. Snow, *The Two Cultures and the Scientific Revolution*, and Mazzio, "Shakespeare and Science." Influential studies of early modern literature and science include Howard Marchitello, *The Machine in the Text*, 185–213; Elizabeth Spiller, *Science, Reading, and Renaissance Literature*; Henry S. Turner, *The English Renaissance Stage*; Paula Blank, *Shakespeare and the Mismeasure of Renaissance Man*; the essays in *Science, Literature, and Rhetoric in Early Modern England*, ed. Cummins and Burchell; Mary T. Crane, *Losing Touch with Nature*; Katherine Eggert, *Disknowledge*; and *Spectacular Science, Technology, and Superstition in the Age of Shakespeare*, ed. Chiari and Popelard.
54. Crane, *Losing Touch with Nature*, 18, 22–27.
55. See, e.g., *Representation in Scientific Practice*, ed. Lynch and Woolgar.
56. This approach joins recent work in the history of science, especially by Peter Harrison. See, e.g., *The Bible, Protestantism, and the Rise of Natural Science* and *The Fall of Man and the Foundations of Science*, as well as his edited anthology of essays in *The Cambridge Companion to Science and Religion* and "The Book of Nature and Early Modern Science." Alexander Wragge-Morley also addresses relations between theology, metaphysics, and representational practices among natural philosophers of the Royal Society in *Aesthetic Science*. On the debates among members of the Royal Society over the proper relation between theology and natural philosophy, see William Poole, "Francis Lodwick's Creation."
57. Principe, *The Scientific Revolution: A Very Short Introduction*, 57.
58. An important survey of misleading narratives about religion in accounts of science can be found in the essays collected in *Galileo Goes to Jail and Other Myths about Science and Religion*, ed. Numbers.
59. For a valuable overview of developments in the field, see Ken Jackson and Arthur Marotti, "The Turn to Religion in Early Modern English Studies."
60. Shuger, *Habits of Thought in the English Renaissance*, 6.
61. Diehl, *Staging Reform, Reforming the Stage*; Knapp, *Shakespeare's Tribe*.

62. See, e.g., Shuger, *Political Theologies in Shakespeare's England* and the essays collected in *Political Theology and Early Modernity*, ed. Hammill and Lupton.
63. For debates about the secular in early modern drama, see Anthony Dawson and Paul Yachnin, *The Culture of Playgoing in Shakespeare's England.*
64. I do not mean to suggest that Shakespeare's religious background is irrelevant. The 1757 discovery of the "Borromeo will" requesting Catholic last rites in the ceiling of Shakespeare's birthplace with the possible signature of his father John is one piece of evidence suggesting that Shakespeare's family remained defiantly Catholic during his lifetime. For more on his possible Catholic background, see Richard Wilson, *Secret Shakespeare.* For further debate on the authenticity of the will and Shakespeare's religious affiliation, see Jonathan Bate, "The Case for the Folio," and Katherine Duncan-Jones, *Shakespeare: An Ungentle Life.* René Weis offers a middle position in *Shakespeare Revealed* by claiming that Shakespeare's own religious views are not really discernable in his works. See also Jean-Christoph Mayer, *Shakespeare's Hybrid Faith.* Bate echoes this sentiment in *Soul of An Age* where he notes that "Shakespeare's political beliefs are as elusive as his religion, his sexuality, and just about everything else about him that matters" (315).
65. Ndiaye outlines and defends her "global framework" in *Scripts of Blackness*, 15.
66. Eisenstein, *The Printing Revolution in Early Modern Europe.* Johns in *The Nature of the Book* contests Eisenstein's claim that print fixes texts by exploring the realities of piracy and problematic knowledge claims in early modern printing more broadly, especially 10–19.
67. Eisenstein, *The Printing Revolution*, 187–254.
68. Kusukawa, *Picturing the Book of Nature*, 249.
69. See Paula Findlen, "Natural History," 442–43.
70. Vesalius, *De humani corporis fabrica* (1543). Quoted in Long, "Objects of Art / Objects of Nature," 77. Pamela H. Smith also notes that images produced by print should be understood not just in terms of their "proliferation" but also in terms of the enhanced "epistemological status of images in the making of natural knowledge." See "Art, Science, and Visual Culture in Early Modern Europe," 85–6.
71. As N. Katherine Hayles notes in relation to her theory of "infomatics," "information technologies do more than change modes of text production, storage, and dissemination. They fundamentally alter the relation of signified to signifier." See *How We Became Posthuman*, 29–30.
72. See, e.g., Ian Hacking, *Representing and Intervening*, and the anthology of essays on *Representation in Scientific Practice*, ed. Lynch and Woolgar. A more narrow study is Martin Rudwick, "The Emergence of a Visual

Language for Geological Science, 1760–1840." A broader, more philosophical approach to visual epistemology can be found in Rudolf Arnheim, *Visual Thinking*.

73. See Pamela Smith, *The Body of the Artisan* and the essays by Smith and others in the "Focus" section on "Science and Visual Culture" in *Isis*.
74. Long, "Objects of Art / Objects of Nature," 63.
75. The quote is from Mitchell, "Interdisciplinarity and Visual Culture," 540, which draws on his monograph *Iconology: Image, Text, Ideology*. See also his *Picture Theory*. For early modern studies that incorporate Mitchell's more expansive view of visuality, see *Visuality Before and Beyond the Renaissance*, ed. Nelson, and *Early Modern Visual Culture*, ed. Erickson and Hulse. For a different understanding of visual culture, see the essays in *Visual Culture*, ed. Bryson, Holly, and Moxey.
76. For an account of Shakespeare's relationship with visual arts, see Astington *Stage and Picture in the English Renaissance*. See also Frederick Kiefer, *Shakespeare's Visual Theater*. For a more nuanced view of the emblematic nature of Shakespeare's imagery in relation to narrative structures and mimesis, see Judith Dundas, "Shakespeare's Imagery."
77. For example, Ong identifies print technology as a "reduction to special form" that "fixes everything" in *Ramus, Method, and the Decay of Dialogue*, 109. See also McLuhan, *The Gutenberg Galaxy*, 61–93.
78. See Benjamin, "Work of Art," 229–33.
79. As Brian Copenhaver notes, "Through the sixteenth century, the new technology of printing strengthened the partnership [between images and nature] by multiplying, stabilizing, and disseminating images on the printed page." Such proliferation made both new learning and new "art" "more believable." See "Magic," 527.
80. In *The Advancement of Learning*, for example, Bacon links together "empirics and old women." See Kiernan, ed. 101.

CHAPTER 2

Representing Nature

This chapter provides a historical and conceptual foundation for subsequent chapters in the book by analyzing new ways of understanding and representing nature in early modern Europe. I focus in particular on the transnational technology of moveable type as it helped to frame nature as a source and subject of visual representation, shifting relations between words, images, and objects of knowledge in the process. Surveying a range of books printed in early modern England and elsewhere in Europe, I note that the divine order of nature was redefined in the period in response to the growing power of print as a mode of visual and textual representation. Earlier accounts of the Book of Nature as a passive emblem of God's creation gave way in this period to more dynamic efforts to explore nature itself in and through hermeneutic systems and new logics. For example, the widely popular herbals of William Turner and John Gerard helped to establish print as a new mode of mirroring nature by remapping relations between text and image. Shakespeare's portrait of Friar Laurence in *Romeo and Juliet* reflects this historical reframing of nature, as do printed editions of Pliny and a range of books of similar works printed in England and elsewhere.

A. Kitch, *Shakespeare's Theater of Nature*, Palgrave Studies in Literature, Science and Medicine,
https://doi.org/10.1007/978-3-031-78082-0_2

The Grammar of Nature

In its patristic and medieval iterations, the Book of Nature followed the logic of Platonic idealism as a model of inscription. As we saw in the Introduction, Augustine provided the fullest defense of the trope in his commentary on the Psalms, which urges Christians to "roam through the whole creation" in order to see how "everything cries out to you of its Author."[1] The argument that God writes laws with his fingers overlaps with the argument that the created world is a "book" authored by God, which parallels the "word" made "flesh" through the Incarnation of Christ. As examples from Paul, Augustine, and others illustrate, the Book of Nature develops as a structure of allegory, analogy, and visual incarnation by which visible phenomena can be understood in relation invisible essences.

The Book of Nature not only expresses the divine potential of the *ordo creationis*, it also depends on a theory of language in which signifier and signified unite under the umbrella of a stable and transhistorical sign. But acts of reading the Book of Nature as an expression of God's creation gave way in early modern Europe to new efforts to understand nature itself in and through hermeneutic systems based on different methods and logics. In describing his discovery of the circulation of the blood in the human body for the first time, for example, the celebrated natural philosopher William Harvey proclaimed that he gained his knowledge "not from books but from dissections, not from the tenets of Philosophers but from the fabric of Nature."[2] Harvey implored readers of his *Anatomical Exercitations* (1628) to be "instructed" by their "own eyes" and to make "ocular inspection" of "things themselves," yet he also retained the older logic of the Book of Nature in his effort to understand "Nature's Book," which is "so open, and legible."[3] His use of the word "dissection," which had only recently entered the English language, registers the nascent science of anatomy in early modern England following the incorporation of the Barber-Surgeons Company in 1540, which conducted annual dissections attended by surgeons.[4] As we will see in later sections of this book, such early modern dissection mixed old and new epistemologies together by framing the direct observation of the body in terms of the theological presence of the *ordo creationis*.

Like Harvey, the clergyman Thomas Tymme promises in his *Dialogue Philosophical* (1612) to open the "secret closet" of "nature," reading "Nature's book" so as to contemplate the "works of the great and incomprehensible God." Observing the motion and beauty of planetary movement, for example, leads Tymme to praise the "power and wisdom of the omnipotent Creator." But the best way to behold God, he avows, is in the "little world" of human beings who have been "imprinted" with a "Divine Essence." For both Tymme and Harvey, the study of the self is a gateway to forms of knowledge about the greater universe. It is only through the contemplation of the microcosm of the individual human that we can gain access to the "Universal" elements.[5] This focus on the value of the microcosm was not unusual, but, unlike earlier accounts, Tymme frames the order of nature as a source of knowledge and creative potential in its own right, understanding representation and discovery as interconnected elements of knowledge production.[6]

The "book" part of the Book of Nature also resonated with Protestants like Tymme who sought a return to the original word of God. Incorporating the bookishness of the Jews as people of the book and the "testament" of the Hebraic scriptures, not to mention traditions of Jewish textual exegesis, Christian scripturalism for early modern Protestants elevated the book as an emblem of authority and completeness. As the German philologist Ernst Robert Curtius suggests in his classic survey of "The Book as Symbol," "it was through Christianity that the book received its highest consecration."[7] In the historically complex and contested shift from the holy scroll of the Torah to the codex of the bible, the Book of Nature emerges as a principle of order and legibility that governs relations between God and mankind, between the order of created things (*res*) and written and spoken language (*verba*).

Early modern natural philosophers also used new techniques of print production to render the expansive order of creation known and visible. Print promised to transform the unending diversity of birds, rocks, flowers, trees, animals, and celestial objects into legible words and images. The fixed covers of a book's leather or vellum binding also provided tangible evidence of such nature, just as the rectangular pages of the codex with its invitation to ordered production of textuality and its recursive ability to move both forward and backward produced new forms of visibility and knowledge. As Curtius notes, scriptural metaphors are extensive in both the Old and New Testaments, from the claim that Mosaic Law is "written with the finger of God" (Exodus 31:18) to the declaration of the Psalmist that "my tongue is the pen of a ready writer" (Ps. 45:1).[8] Yet such examples neither invoke nor follow the logic of the Book of Nature, which develops a particular logic of how to know and organize the signs of divine creation.

Some early modern natural philosophers developed this logic in new ways. In the first book of the *Advancement of Learning*, for example, Bacon rebuked those who "disdain to spell and so by degrees to read in the volume of God's works."[9] Bacon's disciple Robert Boyle likewise sought to read "the stenography of God's omniscient hand" in his scientific experiments.[10] Such examples remind us that the order of nature remained enmeshed in figurative language well into the seventeenth century. Bacon tried to understand the Book of Nature as part of his effort to recover the fullness of understanding nature and the clarity of original language in the Garden of Eden, but he did so with a quite different sense of the language of nature.[11] Bacon was skeptical about human capacity to discern the presence of God in the natural world, urging a more humble process of careful observation of particulars in order to work inductively toward larger laws.[12] But he also understood the scientific enterprise as part of an eschatological project. In his late and unfinished manuscript *Valerius Terminus, or The Interpretation of Nature*, for example, he claimed that the central goal of science was the "restitution" of human sovereignty over nature once enjoyed by Adam and Eve in Eden.[13] In surveying the order of nature for signs of its divine creation in his *New Organum*, moreover, Bacon also sought to make "experience" itself "literate."[14] By learning its "letters," Bacon hoped, such natural experience could be drawn out of the particular and private concerns of individuals and into the public sphere, where it could generate new forms of human understanding. In the process, Bacon rejected any stark division between art and nature, arguing in the second book of *The Advancement of Learning* that "it is the duty of Art of perfect and exalt Nature."[15] Bacon also argued that mechanical history or the history of "nature wrought" should be studied as part of natural history since "nature cannot appear so fully in the liberty of Nature, as in the trials and vexations of Art."[16] A central feature of his effort to reform knowledge was to transform the medieval "experiment" as a spontaneous and private event into a more ordered and organized "experiment" that would be both "literate" and public.[17]

In one sense, of course, these efforts by Bacon and other natural philosophers to "read in the volume of God's works" were not new. The Carolingian Archbishop Rabanas Maurus had made a similar attempt as far back as the ninth century CE in his poetic description of the Book of Nature:

> For the fingers rejoice in writing, the eyes in seeing,
> And the mind unrolls the words of God in their hidden sense.
> No work rises up that aged antiquity does not overcome

Or that wicked time does not overturn.
Letters alone escape ruin and ward off death,
Letters alone in books renew the past.
Indeed God's finger carved letters on suitable rock
When He gave the law to his people;
These letters disclose all that is in the world,
Has been, and may chance appear in the future.[18]

For Maurus, who also composed a study of grammar and an encyclopedia "On the Nature of Things" (*De rerum naturis*), the art of writing (*grammata*) both "disclose[s] all that is in the world" and transcends that mutable world of time and mortality. His allusion to the tablets of Moses reveals the biblical foundation of his verse, but letters for Maurus also have the capacity to "renew the past." Such an expansive understanding of the power of language relies on the formal discipline of *grammatica*, which institutionalized the study of language, grammar, and textual criticism and governed relations between readers, texts, writers, and communities throughout Europe. Initially focused on the works of Virgil in the centuries after the classical poet's death, *grammatica* became the primary engine of Christian theology and commentary for more than a thousand years beginning in the fourth century CE. Unlike its sister disciplines of rhetoric (*rhetorica*) and dialectics (*dialectica*), *grammatica* was a science of interpretation that included the rules for reading and oral citation (*lectio*), guidelines for interpreting and commenting on texts (*enarratio*), and rules for correcting orthography and grammar to establish authoritative editions of texts (*emendatio*). As the foundational discipline of the liberal arts, *grammatica* offered a primary lens for understanding how to gain knowledge about the order of nature.

In his comprehensive survey of *grammatica* as both an art and a science, Martin Irvine notes that it

> was not simply descriptive, isolating and labelling the parts of its subject matter, but *productive* of knowledge: it supplied a network of presuppositions, discursive strategies, and rules for argumentation that governed inquiry and provided the grounds of possibility for knowledge. *Grammatica* was thus a paradigm, in Kuhn's sense of a conventional and consensual epistemic model acknowledged in varying degrees of self-consciousness by individual practitioners of a discipline, and a discursive practice that supplied the conditions for knowledge, providing the discursive means for constituting textualized linguistic objects as objects of knowledge *per se*.[19]

In describing *grammatica* as a paradigm, Irving links the discipline to the history of science following Kuhn's influential model. As a discursive practice, moreover, *grammatica* was foundational for many different disciplines of knowledge. It gave its practitioners methods for organizing links between signifying chains of words and the objects they signified. Isidore of Seville's *Etymologies* (*Etymologiae*, c. 625 CE) was one of the most influential examples of such organization for more than a thousand years, defining key words such as "nature" and "birth" by situating them in relation to vast etymological networks.[20] Efforts to study plants and animals by early modern natural historians drew on similar principles to define the expansive order of nature using a method that Brian W. Ogilvie calls a "science of describing" whose aim was to generate "a comprehensive catalogue of nature."[21] But new approaches to such describing also challenged traditional understandings of these networks in early modern Europe. In the early seventeenth century, for example, Galileo redeployed both the art of *grammatica* and the trope of the Book of Nature by claiming in the *Assayer* (1623) that

> philosophy is written in this grand book, the universe, which stands continually open to our gaze. But the book cannot be understood unless one first learns to comprehend the language and to read the letters in which it is composed. It is written in the language of mathematics, and its characters are triangles, circles, and other geometric figures, without which it is humanly impossible to understand a single word of it.[22]

Galileo's version of the Book of Nature transforms *grammatica* by substituting "triangles, circles, and other geometric figures" for the textuality at the discipline's heart. Galileo indeed regards such abstract numerals as the hieroglyphics of nature itself.[23] After all, the claim that nature itself is written in a language of mathematics was as old as Plato, but Galileo frames his discoveries through *grammatica* in order to define nature itself as a form of language.[24] Galileo's notion of mathematics as the "language" of nature goes beyond the modes of textuality associated with *grammatica*, signaling its increasing irrelevance in empirical science over the course of later centuries.[25] Like Bacon and Boyle, he elevated the Book of Nature as a model for reading the order of nature over the invisible divinity that created it. As he explained in a letter to the Grand Duchess Christina in 1615, for example, when the bible and the Book of Nature contradict one another, the latter must take precedence.

Galileo's revision of the Book of Nature also depended on his use of new visual technologies, namely the telescope or "perspective glass" through which he observed the cosmos. And as he trained his perspective glass on what he called the "things themselves," he uncovered representational structures within nature.[26] In a letter that alludes to his discovery of sun spots, for example, Galileo asserts that he has "no doubt that they are real objects and not mere appearances or illusions in the eyes or of the lenses of the telescope."[27] In order to prove their reality, however, he relied on two analogies: (1) the similarity of the sun spots to the shadows in the moon caused by lunar mountains that he had already observed through his telescope and (2) the material link between the spots on the sun and ink "stamped" (*stampata*) on white paper. As Howard Marchitello notes, these analogies imitate the mediation of the telescope itself as a visual technology:

> Galileo's sunspots do not simply emerge from the surface of the sun into human understanding as the result of the observation of nature, but rather from the very technology that enables such observation in the first place. Sunspots, we can say, were invented by the collaboration of nature and artifice under the direction of Galileo who manipulates both the telescope and the natural light it captures and employs both in the service of a technology of representation devised in order to make the sunspots manifest to the scientist and subsequently—through the technologies of illustration and print—to his readers.[28]

As Marchitello argues, Galileo's telescope was not just an important new visual technology that enabled scientific discovery—it also revealed nature itself as a structure of representation. Galileo both located a new language of the Book of Nature in mathematics and reconfigured the use and value of languages of nature themselves, with potentially paradoxical results. His metaphor of imprinting ink on paper to explain sunspots replicates the revolutionary process that he employed to transfer the visual data from the telescope to sheets of paper. As he explained in a letter about his process of discovery, he pointed the telescope toward the sun as if he were going to look at it directly but then placed a "flat white sheet of paper about a foot from the concave lens" onto which the sun disk would be illuminated, in essence expanding the image of the sun in order to see the spots by projecting the image on paper. In other words, the physical process of projecting the image of the sunspots produced the sunspots themselves. The image projected on the paper was the prototype for the image produced in the printed edition of his *Discourse on Floating Bodies* (*Discorso intorno alle cose che stanno in sù l'acqua*, 1612), which sold out almost immediately.[29] This process of transcription and dissemination provides a famous example of how printing not only helped to generate new knowledge about nature, but also reframed nature as an object of representation.

Refashioning Natural History

Galileo's discoveries reached wide audiences through the new medium of print, following earlier examples of printed books of nature that had circulated in Europe for at least a century. Such products of the press both brought the order of nature closer to readers and reconceptualized that order in new ways. One of the most widely disseminated examples of such a book of nature was Pliny's *Natural History* (*Naturalis historia*, c. 79 CE), which remained in circulation in both manuscript and printed copies from the time of his death through the eighteenth century. Pliny's encyclopedia defined nature as the ultimate manifestation of God's presence and power. As a compilation of textual authorities including Aristotle, Varro, Ovid, Flaccus, and others outside of the Latin tradition, it reflected the interconnectedness of the order of nature and its textual representation, emphasizing how the different elements of the natural world relied on and influenced one another. This interconnectedness, for Pliny, was ultimately a reflection of the divine interconnectedness and divine providence that governed all aspects of existence.

Like Christian natural philosophers, moreover, Pliny in Stoic fashion sought to deepen the understanding and appreciation of nature's relationship with God, encouraging his readers to marvel at the natural world and recognize the divine presence within it. Moreover, unlike other classical authors who were widely read in early modern Europe, Pliny did not need to be rediscovered by humanists for the simple reason that he was never lost.[30] Following the discovery of printing in Europe, however, Pliny's encyclopedia became even more popular, and this enhanced popularity reveals much about the way that "nature" was understood in the period.[31] Humanists may not have needed to rediscover Pliny, but they redefined his account of "nature" in significant ways during the fifteenth and sixteenth centuries. Scholars were especially interested in Pliny's range of Latin terms to describe the natural world, which they cross referenced with newly discovered Greek lexicons. The most important early modern commentaries prepared by Beatus Rhenanus, Sigismund Gelenius, and

Ermolao Barbaro, among others, were dedicated to establishing a new and more stable text for circulation. As Charles Nauert remarks, "all of this scholarly work on Pliny was humanistic in a very specific sense of the term: the humanists were dealing with the Natural History as a classical text, applying their knowledge of manuscripts—and perhaps of classical Latin grammar, ancient science, and other ancient authors—to the improvement of that text."[32] These efforts would seem to confirm Eisenstein's claim that printing over multiple editions allows a text to be stabilized in a way that manuscript does not. Printed copies allowed additional commentators to compare one text with another and to incorporate manuscript copies and commentaries into new and expanded editions that, in turn, allowed other editors to do the same.[33]

Over the course of the sixteenth century, however, debates about textual authorities and terminology among humanist editors of Pliny focused on modes of direct observation of the order of nature that might enhance and correct Pliny's claims through systematic observation of the natural world rather than by relying on textual authority. While some later historians dismissed Pliny as an armchair scientist and lowly compiler of the knowledge produced by other authors, other scholars regarded his commitment to detailed facts about the natural world as a self-conscious rejection of the kind of theorizing around nature that he found in examples such as Lucretius and Seneca.[34] Pliny also followed a clear hierarchy of forms of knowledge as he outlines in his overview in the first book, which moves from descriptions of the four elements and the cosmos through geography to humans, then to animals, plants, trees, herbs, metals, and stones. This movement from macrocosm to microcosm and beyond to the world of objects situates humans at the center of creation while also suggesting a degree of continuity between the various levels of the order of nature.

Like Ovid's *Metamorphoses* (c. 8 CE), Pliny's *Natural History* was a staple of secondary schools in England such as the King's New School that Shakespeare attended in Stratford.[35] Both classical texts were also subject to forms of allegorical reinterpretation that pedagogues hoped would make them more palatable to sixteenth-century students and their parents. Although Pliny's compilation of plant knowledge appears to be a straightforward accumulation of facts, he was also an avowed Stoic. This Stoicism, as Aude Doody notes, encouraged a view of "Nature both as a quasi-divine creative force and as manifest and immanent in each natural thing."[36] Early modern printed editions highlighted this divine immanence while translating it into a more overtly Christian context. The earliest printed edition of Pliny in England was an abridged summary printed in 1566 by Henry Denham and translated from a French text by "I.A." In the preface to the reader, the translator remarks that Pliny reveals how

> the works of God are marvelous, not only in us his creatures, whom he hath fashioned and formed like to his similitude, but also in others, as beasts, fouls, fishes, trees, plants, and such like, whose miraculous works, although unto us some things seem incredible, yet if we did consider the omnipotency of God, unto whom nothing is impossible, doubtless we should not run into so many dangers of sin as we daily do.[37]

The first full translation of Pliny into English was undertaken by the clergyman, physician, and classical scholar Philemon Holland, whose two-volume folio of Pliny reached London bookstalls beginning in 1601. This triumphant marriage of humanist scholarship and print production followed earlier editions of Pliny in Europe (beginning with the Venetian edition of 1568) that conjoined print technology, natural philosophy, and theology in new ways. Abridgements of Pliny appeared in England under the title of *The Secrets and Wonders of the World* in 1585 and 1592. In his preface, Holland addresses a potential controversy that the non-Christian Pliny gives too much agency to Nature over God. But Holland reassures the reader that he has consulted "sundry divines" on the subject who assure compatibility between Pliny with Christian doctrines. He also includes a letter from a patron that suggests that Pliny, even though pagan, understands nature to be a necessary "veil" on the "glorious light" of God that would otherwise blind us if we looked at it directly.[38]

Holland also frames his translation in explicitly theatrical language. He describes his earlier translation of Livy, printed in 1600, as a Greek "interlude" or theatrical performance that was both applauded and confirmed with silence. Having thus shown himself on the "stage," he presumes to make another attempt at performance through print production.[39] There is no evidence that Holland was a fixture at London playhouses, but we do know that at least one author of English theater was a fan: Shakespeare likely drew on Holland's translation of Pliny in composing Othello's reply to Iago's manipulative counsel of "patience" in act 3.[40] But Shakespeare's indebtedness to Pliny went beyond this particular quotation. Pliny's encyclopedic encompassing of natural, supernatural, human, animal, and material forces resonates with Shakespeare's own representation of the order of nature in his plays.

Consider, for example, one of Shakespeare's most beloved characters (and most enthusiastic gardeners): Friar Laurence in *Romeo and Juliet*. We first meet the friar in act 2 as he surveys his monastic garden at dawn:

> O, mickle is the powerful grace that lies
> In plants, herbs, stones, and their true qualities,
> For naught so vile that on the earth doth live,
> But to the earth some special good doth give;
> Nor aught so good but, strained from that fair use,
> Revolts from true birth, stumbling on abuse.
> Virtue itself turns vice, being misapplied,
> And vice sometimes by action dignified.
> Within the infant rind of this small flower
> Poison hath residence and medicine power:
> For this, being smelt, with that part cheers each part;
> Being tasted, slays all senses with the heart.
> Two such opposed kings encamp them still
> In man as well as herbs, grace and rude will;
> And where the worser is predominant,
> Full soon the canker death eats up that plant. (2.2.15–30)

Like Pliny, Friar Laurence understands the order of nature as a source of moral and practical knowledge.[41] Although he also alludes to the medicinal properties of plants, he is more interested in outlining a pattern of good and evil in the order of nature that includes both plants and humans.[42] The divine "grace" in created matter can turn evil, he notes, if "misapplied." This claim echoes the Platonic notion of the *pharmakon*, which united "poison" and "medicine" in the natural world.[43] But Friar Laurence also outlines a hermeneutics of grace that links the *ordo creationis* to his medical practices and his ethical worldview, however problematic that worldview turns out to be. Plinian natural history becomes a source of a complex cultural and philosophical account of creation on the Shakespearean stage.

In praising the "secret qualities" and "hidden virtues" of plants, herbs, and stones, Friar Laurence locates divinity in even the smallest flower. The actor who played the role of Friar Laurence in original performances at the Theater in Shoreditch might well have carried specimens of such nature directly into the playhouse in the form of wildflowers and herbs plucked from the nearby Finsbury fields.[44] Such living matter would have stood in stark contrast to the entirely manmade space of the theater itself, which enclosed the audience in a completely artificial environment. But Friar

Laurence also joins a long line of historical figures attempting to discern the secrets of nature, including Pliny and Cornelius Agrippa's *Three Books of Occult Philosophy* (*De Occulta Philosophia libri III*, 1533). Despite his status as a natural magician, Agrippa follows Pliny in understanding nature and seeking to bring "powers of all things natural and celestial [...] lying hidden in nature into the open."[45]

In his theological approach to reading the Book of Nature, Friar Laurence also invokes Reformation debates about the nature of God. Despite his status as a Franciscan friar in Catholic Italy, Friar Laurence takes a rather Protestant view of original sin. His language echoes, in fact, Martin Luther's observation that human nature is "infected or poisoned" and "altogether corrupt."[46] Closer to England, Friar Laurence also resembles the Protestant translator and martyr William Tyndale, who describes the "poison inclination & corrupt nature" in which we are "conceived and born" in his revolutionary translation of the New Testament.[47] Such language was also popular in medieval and early modern herbals that remained a dominant source of natural knowledge well into the seventeenth century. In fact, Friar Laurence sounds a lot like the famous English botanist William Turner, who studied the order of nature for its medicinal value and claimed that plants revealed God as the "author of all goodness."[48] Like Friar Laurence, for example, Turner mixed his account of the virtues and plants with references to their capacity to poison humans. Friar Laurence also echoes the notorious botanist, astrologist, and physician Nicolas Culpepper, who published recipes for herbal remedies and praised the image of God he found "stamped" on every blade of grass.[49] Similar claims appear in the *Herbal* (1597) of John Gerard, which supplanted Turner's *New Herbal* to become the most popular herbal in England for almost a century.

Both Turner and Gerard emphasized direct observation of actual specimens of plants on the model of continental herbalists such as Otto Brunfels in Germany, whose *Images of Living Plants* (*Herbarum vivae eicones*, 1530–36) aimed to reproduce accurate images of actual works rather than relying on existing illustrations in Pliny and other classical sources.[50] Shakespeare and Gerard traveled in the same circles, since both men shared a patron in Lord Burghley.[51] A later, expanded edition of Gerard's herbal grew to more than 1600 pages and was published by the London apothecary Thomas Johnson in 1633. Its newly designed frontispiece offers a visual landscape linking the genre to the order of nature as a structure of divine representation (Fig. 2.1):

Fig. 2.1 Illuminated title page of the 1633 edition of John Gerard's *Herbal, or General History of Plants* (1597), engraved by John Payne. RB 485990, The Huntington Library, San Marino, CA

The elaborate engraving by John Payne frames Gerard's popular herbal in terms of both classical and Christian iconography. The Roman goddess of plants Ceres occupies the top left corner, just above the sculpted figure of the ancient Greek botanist Theophrastus. The Roman goddess of fruit trees Pomona stands in the top right panel, immediately above a depiction of the Greek pharmacologist Pedanius Dioscorides. (Some scholars have speculated without compelling evidence that the figure of Dioscorides was modeled on Shakespeare.[52]) Beneath the portrait of Dioscorides, Payne represents a cultivated garden of trees, plants, and herbs that may refer to Gerard's own celebrated garden. The scroll near the garden contains a Latin sentence from Genesis: "Ecce dedi vobis omnes herbas sementantes semen, quæ sunt" ("Behold, I gave you all the plants that exist and bear seeds") (Genesis 1:29). Below this image a different scroll contains an elegiac couplet that proclaims "Excideret ne tibi divini mumeris Author; / Præsentem monstrat quælibet: herba Deum" ("Lest the Author of the divine work might mislead you / Any plant will itself reveal the presence of God").

In the dedication of his herbal to the powerful Lord Treasurer William Cecil, Gerard praises the power of plants to draw humans toward the contemplation of the divine and suggests that the variety of color in nature outdoes any painter or jeweler. Citing Pliny as a primary example, Gerard notes that even brute beasts have some knowledge of the hidden "virtues" of plants as a source of nutrition and medical treatment. (He observes, for example, that "dyers" and other artists derive their art from the study of such nature.) His herbal offers practical advice about plants as sources of food and medicine while also participating in the expansive representational capacity of nature as a structure of representation.[53] The frontispiece provides an emblematic representation of this capacity, which centers on a divine hermeneutics that seeks the hidden virtues of all created matter, while the text itself also produces the order of nature as a system of verbal signs that translates the natural order into language. Like Friar Laurence, readers of Gerard and Pliny must learn to bridge the gap between signifier and signified, a hermeneutic lesson that draws on a long tradition of locating God's presence in the world.

A Treasury of Signs

The efforts of Pliny, early modern botanists, and the fictional Friar Laurence to read the Book of Nature recall Michel Foucault's description of the premodern "episteme" in *The Order of Things* (1966) whereby the "the great metaphor of the book [...] forces language to reside in the world, among the plants, the herbs, the stones, and the animals."[54] For Foucault, the order of nature makes itself known through a "treasury of signs" that demand careful interpretation. Such signs include both natural and artificial creations—both the medical practices of Paracelsus, for example, and the painting *Las Meninas* (1656) by Velazquez.[55] Though Foucault's methodology has been criticized for its overly rigid classification of historical periods (or "epistemes") and for its misreading of some primary sources, his account of the hermeneutics of nature remains influential.[56] Yet Foucault adopts a largely secular focus that overlooks important theological foundations of the material he analyzes.

The Tetragrammaton featured prominently at the top and center of the frontispiece of Gerard's herbal provides clear evidence of such foundations. The popular device derives from a passage in Exodus when Moses asks God to name himself, receiving only the enigmatic response "I AM THAT I AM" (3:13). In Hebrew, this passage appears as a series of unpronounceable letters, "יהוה" or "YHWH," that reflect the Hebrew verb "to be" ("היה"). In naming God, in other words, the Tetragrammaton violates grammatical norms to recognize the limits of representing the divine in language.[57] Far from illustrating what Derrida calls "logocentrism," this Tetragrammaton in fact figures a foundational dialectic of absence and presence.[58] New Testament revisions of the Tetragrammaton sought firmer epistemological footing for the relation of God to the order of nature, especially in the doctrine of Incarnation. As the opening lines of the gospel of John declare: "In the beginning was the Word, and the Word was with God, and the Word was God" (1:1). Such a claim asserts the linguistic presence of Christ, tracing movement from "with" to "was," from language as contiguous with God toward language expressing the essence of that divinity.

Ontologically, the letter could incarnate the body of Christ, even if it remained secondary to the primary and pre-existent Being of God.[59] Many Christians therefore regarded the alphabet itself as divinely structured. Paul echoes such logic of divine inscription even as the medium of writing shifts from stone to flesh. Christ achieves living presence through embodied life on earth, but his presence after his death lives on in the written word and in the hearts of his followers just as his spirit returns to God. To understand God's Incarnation as an act of textual inscription was therefore to recognize how the bible both reveals and conceals its divine origins. In his Greek homilies on Leviticus, for example, Origen notes that few detected the concealed divinity of Christ in his own lifetime, just as readers of the divine text may struggle to interpret the divine presence concealed by the words. For Origen, the Word of God was "covered by the veil of flesh," which conceals the "spiritual sense within." Readers therefore had to achieve new spiritual enlightenment in order to "see divine spirit through the letter's veil."[60] Others, such as Jerome, interpreted the sacrament of divine communion as a form of textual incarnation: "when we are listening to the word of God," he noted, "God's Word and Christ's flesh and blood are being poured into our ears." Christ's request to his followers that they consume his flesh and blood was therefore part of the same process in which Christ was incarnated in the sermons and scriptures to be "digested" by readers and listeners.[61]

Paul's first convert to Christianity in Athens, writing under the pseudonym Denys the Areopagite, held a quite different view on the capacity of language to capture divinity.[62] His *Divine Names* (*Περὶ θείων ὀνομάτων*, c. 530 CE) argued that God was both beyond language and above being. For Denys, the words of the Holy Trinity, "Father," "Son," and "Spirit" provided some access to the nature of his own divinity, but they could not capture his essence because "He is not one of the things that are." In other words, God "is not understood, nothing can be said of Him, He cannot be named."[63] Resisting the logic of the incarnation in John, language for Denys fails to reveal divine being in part because God is not a member of the group of things that merely exist. His ontology is beyond human capacity to know and represent. As a modern editor of Augustine puts it, God is an ineffable and transcendent being who is "known better in not being known," given that it is "easier to say what He is not than what He is."[64]

For Augustine, moreover, the Word of God does not unfold in mortal time. Our fallen, corporeal selves can comprehend God in terms of the three persons, each with a distinct name, according to the necessities of difference, even though the unattainable meaning of God is unified but beyond our apprehension.[65] Yet Augustine proposes the "Book" as a powerful and necessary intermediary between the self and God, one that directs carnal vision toward the spirit of God as creator. The written book for Augustine has the capacity to bond humans to God literally through prayer, meditation, and conversation, while figuratively it does so through sacraments, confession, and the Holy Trinity signaling the unity of God and the soul. For Augustine, therefore, all Christians can access the true image of divinity by reading the bible. Augustine also argues that God can be understood in and through our visual relationship to the created world.[66] He outlines three modes of vision—corporeal, spiritual, and intellectual—that define all successful acts of vision. Corporeal vision enables the literal reading of words on the page and objects in the world, while spiritual vision allows us to imagine actual objects and people who are not present. Intellectual vision generates images of timeless truths such as divine love that do not have literal analogues in the world. Drawing on Platonic accounts of sight, Augustine's three levels of vision combine to make up a unified system that bridges the visible world with its divine creator. Corporeal sight, which operates by the letter, as it were, is reconciled with spiritual and intellectual vision through the language of the bible, and all three are appropriate for the interdependent model of the Book of Nature that Augustine imagines.[67]

The Book of Nature for Augustine thus subordinates the given world to its invisible origins, but his writing also foregrounds questions about the capacity of matter to realize God's presence. In sharp contrast to Augustine's theory of language as participating in the *ordo creationis*, Thomas Aquinas reconceptualized the gap between invisible and visible worlds. As part of his effort to synthesize reason and faith, Aquinas developed new analogies between God and the order of nature. The analogy of attribution, for example, allows God to transmit elements of Himself while acknowledging the limits of our capacity to understand his divine essence. The analogy of proper proportionality, by contrast, concerns a mode of being that is internally different but externally identical, as in the example of man's being, which is different inherently from God's being and also internally different from God.[68] Analogies as Aquinas understood them were not heuristic and therefore could not serve as a foundation of Christian metaphysics, but they nevertheless depend on the representational capacity of language.

The capacity of language to mirror nature became a renewed subject of debate in late medieval Europe. In the great manuscript encyclopedias of nature that came into prominence in the thirteenth century, for example, one finds an abiding tension between different models of language and knowledge, especially in the difference between the figurative and metaphorical approach to language and natural philosophy taken by Augustine and the more literal, objective approach to the natural world taken by Aristotle. The European rediscovery of Aristotelian texts through Arabic translations and commentaries beginning in the twelfth century elevated this hermeneutic tension, which became manifest in the great projects in the thirteenth century of accumulated knowledge known as *florilegia* ("gatherings of flowers") that later centuries called "encyclopedias." Such vast textual mirrors of nature sought to produce comprehensive and syncretic systems of information modeled on scripture.

One of the most influential examples of the genre was the *Speculum maius* or "Great Mirror" (c. 1260 CE) composed by the Dominican friar Vincent Beauvais. Following Dominican tradition, Vincent saw his project as supporting the efforts of Aquinas to synthesize Christianity and Aristotelian philosophy, but his unfinished encyclopedia also contains an Augustinian theory of language as *figura* or veil that covers invisible divinity. The *Speculum maius* was divided into three central parts: (1) a compilation of Greek, Latin, Hebraic, and Islamic accounts of the creation of the world (*Speculum naturale*) in 3718 chapters covering astronomy, psychology, geography, and natural history, (2) treatises on verbal and mechanical arts, the education of princes, medicine, and theology divided into 2374 chapters (*Speculum doctrinale*), and (3) a history of the world from creation to Vincent's own day, mixing theological and secular sources and topics. (A fourth section on moral philosophy called *Speculum morale* was planned but never completed.) Vincent grounds his encyclopedia throughout in biblical texts, even while he weaves together a range of sources, including Pliny's *Natural History*.

Vincent styled himself a collector rather than an author, noting that he made the decision to call the work a "*Speculum* because whatever it contains that is worthy admiring or imitating among the things that have been done or said, in the world visible and invisible, from the beginning down to the present, I have been able by observation to collect from almost innumerable books, and to bring together in brief compass in this one book."[69] Vincent's vast compilation of knowledge asserts the divine origins of the created universe and situates readers within the divine order of nature. In synthesizing Christian, pagan, Egyptian, and Hebraic sources, however, he abstracts them from their historical and cultural contexts in order to designate them as compatible accounts of the order of nature. Such synthesis allows the reader to see beyond the text into its transcendent origins even while deriving specific and sometimes practical knowledge of the order of creation. The *Speculum maius* thus draws on the overlapping meanings of *speculum* in the period—as a spiritual guide, a form of scriptural exegesis, a model for living, and a compendium of divine wisdom.[70] The mirror collapses the external world into text in a way that is intended to inspire self-reflection in the reader. Vincent's text reflects Augustine's definition of the mirror of scripture as a textualization of divine creation into language that is appropriate for humankind's limited capacities of understanding. As Hugh of St. Victor argues, this kind of mirroring "represents our interior image to us, exposing what is beautiful and what deformed in the soul, and how the beauty of justice should be attended to, the beauty of the virtues arranged, and the stains of the vices blotted out."[71]

Vincent's mirroring encyclopedia attempts to dissolve boundaries between text, author, reader, scripture, and the *ordo creationis*. Such mirroring produces both identification and distancing in the reader in a way that parallels Lacan's "mirror stage." The *imago* of the text both completes the fractured self of the devoted reader and instills a break or distance between reader and image as an ego ideal. Christian readers seek to become what they read even if, as Lacan notes, there is a structuring gap or fissure between the subject and image.[72] Vincent anticipates that his readers can see through the text to the invisible order of divinity that structures it. He also imagines the order of nature as a model of self-regulation and self-formation. Acts of mirroring for Vincent and others thus simultaneously reinforce and challenge the separation between reader and word, subject and object. Such acts also demand active engagement and participation by the viewer, who must meditate on the distance between herself and the *ordo creationis* before taking steps to transform herself into a more divine form. This kind of active participation anticipates the way that Jean Bodin and early modern playwrights imagine the *theatrum mundi* as a means of illustrating principles of divinely ordered nature for a broader audience.

A similar process of mirroring occurs in an even more popular spiritual work that circulated broadly in England before the sixteenth century, the *Mirror of the Blessed Life of Jesus Christ* by Nicolas Love. A late example of the medieval *imitatio Christi* genre, the *Mirror* was an English translation of the *Meditationes Vitae Christi* by the Franciscan friar Johannes de Caulibus composed in the early fourteenth century. The editor of the modern English edition estimates that it was more popular in both manuscript and print than almost any other book in the fifteenth century.[73] Love's *Mirror* invites its readers to meditate on the events of Christ's life through reading texts and viewing images. Such images, which Love compares to drama, inspire readers to empathize with Christ's suffering and to recreate new images and narratives based on their own life. Like other works of affective piety, Love's work encourages readers to engage in meditation—to allow their minds to wander freely in order to arrive at new moral and spiritual truths.[74]

Reading meditatively invited those who encountered the *Mirror of the Blessed Life of Jesus Christ* to merge their own subjectivities with Christ, narrowing the distance between readers and their divine object of worship. Such readers internalized and transcended the text, simultaneously finding and losing themselves within a divinely structured order of nature. Love's *imitatio* also had overtly political aims. Scholars have suggested that Love may have endorsed Archbishop Arundel's fifteenth-century campaign against Wycliffe and radical critics of Christianity because *The Mirror* incorporates theological treatises in support of orthodox positions on the eucharist and against biblical translation.[75] Yet the forces critical of Rome against which Love and Arundel defined themselves ultimately triumphed in many northern European countries, producing new genres and attitudes toward selfhood, reading, and the order of nature.

Imprinting the *Ordo Creationis*

The European invention (or discovery) of moveable type, as we have seen, reframed the relation between nature and modes of representation.[76] Close personal relationships between natural philosophers and visual artists who produced the paintings, engravings, drawings, and textual descriptions in printed books also brought printing and visual arts closer together. The brilliantly illustrated books that emerged from this collaboration also shifted the nature of viewing itself. As Lorraine Daston notes, for example, printed images worked to "calibrate the eyes of dispersed observers," enhancing the scope and power of visual observation in the process.[77] William Ashworth also argues that printed illustrations of animals and plants made early modern works of natural philosophy stand apart from classical works by Aristotle and others, which were only textual.[78] These images highlighted the representative capacity of nature as a system of signs.

The widely popular herbals of Turner and Gerard helped to establish print as a new mode of mirroring nature in early modern Europe, especially by remapping relations between text and image. An influential source for these English herbalists, as we have seen, was the *Living Plant Images* of Otto Brunfels, which was published in Strasbourg beginning in 1530. The production process for the book involved more than three hundred illustrations produced by the artist Hans Weiditz (see Fig. 2.2). Brunfels, who was also a medical healer, relied on an informal network of botanists working around Strasburg. But he integrated these living images with classical accounts of plants derived primarily from *De Materia Medica* by Dioscorides.

Shortly after the *Herbarum* was published, Leonhart Fuchs published his *Notable Commentaries on the History of Plants* (*De historia stirpium commentarii insignes*) in Basel in 1542. Working with the draftsmen Heinrich Füllmaurer and Albrecht Meyer, whose designs were brought to

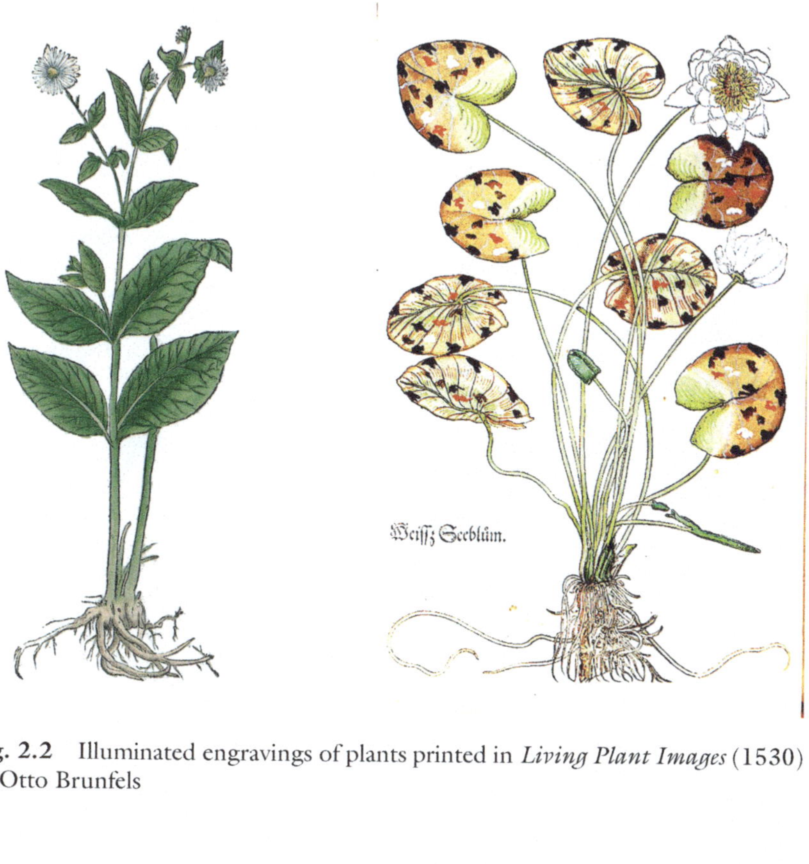

Fig. 2.2 Illuminated engravings of plants printed in *Living Plant Images* (1530) by Otto Brunfels

life by the woodcutter Veit Rudolph Speckle, Fuchs employed illustration as a tool for identifying plant types. He argued that "a picture expresses things more surely and fixes them more deeply in the mind than the bare words of the text" and included even more images than were found in the *Living Plant Images* of Brunfels (see Fig. 2.3). A team of artists engraved the wooden blocks used to create the images and hand-painted them to reproduce them in colors. The book itself highlights this artistic process by including images of the artists involved in this process (see Fig. 2.4).

The hand-colored images of the large plants fill almost the entire folio page of the printed work, magnifying them visually for readers, while the second image celebrates the central role of the image-makers themselves. The two portraits at the top depict the artists Heinrich Füllmaurer and Albert Meyer, while the man at the bottom is the "Sculptor" (i.e., woodcutter) Veit Rudolph Speckle. These innovative herbals of Fuchs and

Fig. 2.3 Illuminated images of plants in the *Notable Commentaries on the History of Plants* of Leonhart Fuchs (1542)

Fig. 2.4 Picturing the visual artists of the *Notable Commentaries on the History of Plants*. The two figures at the top are Heinrich Füllmaurer and Albert Meyer. The lower image pictures the "Sculptor" (i.e., woodcutter) Veit Rudolph Speckle

Brunfels influenced almost all subsequent herbals in Europe, including Gerard's *Herbal*, which includes almost all of the woodcuts from Brunfel. Both of the herbals of Fuchs and Brunfel also include classificatory names with appropriate etymologies; verbal descriptions of stems, leaves, stamens, and flowers; and, most especially, accounts of medical remedies made from the plants included in the book. This recoding of relations between image and word reshaped the form and function of botany as a branch of natural philosophy, lending it new prominence through its capacity for visual representation in print.

English printers also harnessed the power of print to put the order of nature on display in new ways. The earliest printed book to feature images in England was William Caxton's *Mirror of the World* (1481), which provided an encyclopedic catalogue of the wonders of the *ordo creationis.* A condensed translation of a French source, Caxton offers a brief account of God's creation of the world and his act of forming mankind after his own "semblance."[79] In a prologue to the reader, Caxton celebrates the power of the printing press to make the deeds and ideas of the ancients live on "in perpetual memory and remembrance."[80] Following medieval precedents, his encyclopedia integrates divine cosmology with descriptions of plants, animals, weather, and other natural elements around the world. For Caxton, who established a printing press in Bruges before becoming the first commercial printer in England, God created humankind in His image to remind us of our original goodness and to inspire us toward reason (a6). Caxton also describes an original scene of knowledge in which humans contemplate the stars for the first time, coming to understand that they have been created by God and learning that they need to understand that force not directly but rather through his works, just as one understands an architect by studying the buildings that architect has created: "And when the more that they knew of his works & of his wisdom so much more had they the better will to love her creator and maker & to honor him considering that he had made so noble a thing and so worthy as is the heaven in which be the stars that shine bright therein and his other marvelous virtues which they praised much" (sig. b5v).

Before addressing questions of cosmology, botany, geography, and culture, Caxton foregrounds the seven liberal arts, beginning with grammar, which Caxton regards as a foundational tool for all knowledge (sig b8v). To render such textual instruction visible for readers in ways that the text itself cannot illustrate the power of words as tools of scientific knowledge, Caxton includes the first images of the book (see Fig. 2.5).

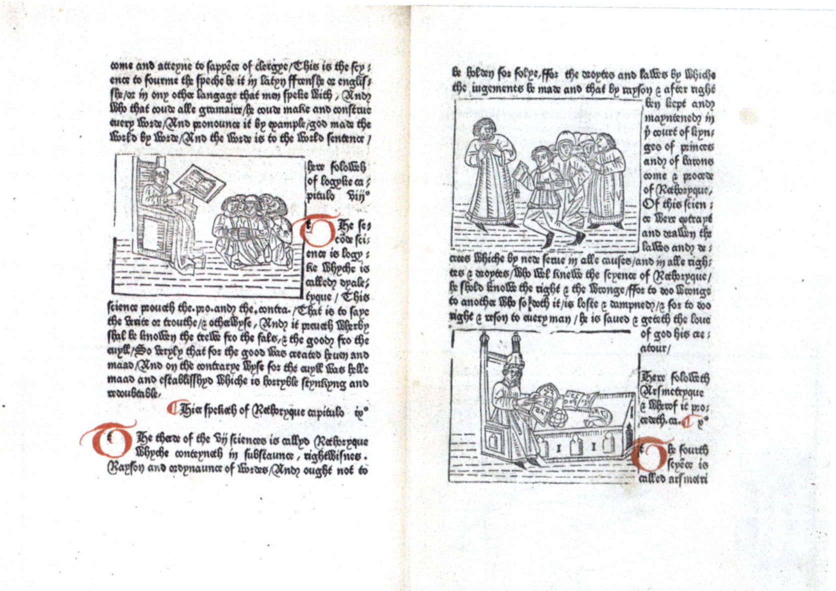

come and atteyne to sapyence of clergye/This is the scyence to fourme the speche be it in latyn ffrensshe or englisshe/or in ony other langage that men speke with, And who that coude alle gramaire/he coude make and construe euery worde/And pronounce it by example/god made the world by worde/And the worde is to the world sentence/

Here foloweth of logyke capitulo viij°

The secode science is logyke whyche is called dyaletyque/This science preueth the.pro.and the.contra./That is to saye the verite or trouthe/& otherwyse, And it preueth wherby shal be knowen the trewe fro the fals,& the good fro the euyll/So veryly that for the good was created heuen and maad/And on the contrarye wyse for the euyll was helle maad and establisshyd whiche is horryble stynkyng and troublable,

Thise speketh of Rethoryque capitulo ix°

The therde of the vij sciences is callyd Rethoryque whyche conteyneth in substaunce, rightwisnes. Rayson and ordynaunce of wordes/And ought not to be holden for folye, ffor the droytes and lawes by whiche the iugements be made and that by rayson & after right ben kept and maynteyned in þ court of kynges of princes and of barons come & procede of Rethoryque, Of this science were extrayt and drawen the lawes and decrees whiche by nede serue in alle causes/and in alle rightes & droytes/who wel knewe the scyence of Rethoryque/he sholde knowe the right & the wronge/ffor to doo wronge to another who so doth it/is loste & dampned/& for to doo right & reson to euery man/he is saued & geteth the loue of god his creatour/

Here foloweth Arsmetryque & wherof it procedeth. ca. x°

The fourth scyence is called arsmetri

Fig. 2.5 Image showing the teaching of grammar and logic in William Caxton's *Mirror of the World* (1481). RB 69795. The Huntington Library, San Marino, CA

Grammar is not just the study of speech for Caxton but the origin of all clerical knowledge because of its capacity to provide access to God. Each successive art or "science," moving from logic to astronomy, builds on but also moves away from this original encounter and experience of divine creation through observation of the world. In his conclusion, Caxton proudly announces that he has achieved his goal of revealing the "inestimable greatness" of the *ordo creationis* (sig. m5).

Caxton modeled his *Mirror of the World* on the *Book of Chronicles* (*Liber chronicarum*, known more frequently as the *Nuremberg Chronicle*) of Hartmann Schedel. First printed by the master printer Anton Koberger in 1493, the *Nuremberg Chronicle* featured woodcuts by Albrecht Dürer, Koberger's godson, who was working at the time in the studio of Michael Wolgemut.[81] The printed chronicle advertises such images as a way to bridge the gap between the reader and the order of nature. As Koberger notes: "I venture to promise you, reader, so great delight in reading it that you will think you are not reading a series of stories, but looking at them with your own eyes."[82] The order of nature for the Nuremberg Chronicle is by no means limited to flora and fauna. It also foregrounds stars and comets, as well as images of European cities, monarchs, and battle scenes. The work grounds this order in the primordial act of divine creation, as pictured in a section on the biblical account of creation in Genesis (see Fig. 2.6). The circle in the image indicates the original nothingness from which God created the world, while God's creative hand in the upper left-hand corner conjures the birds and landscape into life. The spacing and imagery of the printed page illustrate the logic of divine

Fig. 2.6 Illuminated image of the creation of the animals from the *Liber Chronicarum* or "Book of Chronicles" of Hartmann Schedel (1493)

creation itself, capturing images that no longer exist of life on earth before the Fall. The printed book generates a divine order of nature not otherwise visible to readers, making representation on the page the only portal to access nature. There is, indeed, no order of nature without such representation.

Two later examples of printed works produced in England highlight the mimetic capacity of print to reveal the representative structures of nature. Walter Ralegh's *History of the World* (1614) links the discipline of history with the divine order of nature, relying on figurative language in the process. For Ralegh, the God of creation can be known only "by the work of the World" in the form of a "universal Nature" that allows us to perceive the "shadow of his divine Countenance."[83] Acknowledging that God is a "power uneffable" who cannot be "measured," Ralegh nevertheless asserts that we can know Him indirectly by reading "Hieroglyphical Characters" such as the stars and planets that have been "written on these large volumes of the firmament" (2). The frontispiece situates such characters in a symbolic tableau (Fig. 2.7). Clio, the muse of History, hoists the world above her shoulders, a sphere that combines references to biblical locations such as the Garden of Eden with images of early modern ships that highlight Ralegh's own celebrated if controversial voyages to Virginia and Guiana. The omniscient eye of divine providence peers out from above the globe, flanked by the healthy angel of "Good Fame" (*Bona Fama*) on the left and the diseased angel of "Evil Fame" (*Mala Fama*) on the right. But the figures of an aged Experience (*Experientia*) on the left of Clio and a youthful Truth (*Veritas*) on the right indicate the necessary work of reading and observation that is part of history. The four columns moving from left to right extol books as "time's witness" (*Testis Temporus*), hieroglyphs as documents of antiquity (*Nuncia Vetustatis*), rays that symbolize the "light of truth" (*Lux Veritatis*), and climbing ivy heralding the "life of memory" (*Vita Memoriæ*). Although the scheme for these columns derives from a maxim of Cicero quoted by Ben Jonson in the epigraph of the book that praises history as "the teacher of life, the witness of times, the herald of antiquity, the light of truth, and the life of memory," Ralegh's project of universal history reflects early modern tensions between history as a record of divine presence provided by figures of authority and a newer emphasis on direct observation and eyewitness testimony. Like other early modern natural philosophers, Ralegh positions himself against the "rotten ground" of Aristotle's assumptions about the natural world while also acknowledging the limits of the kind of knowledge that humans can gain

Fig. 2.7 Title page of Walter Ralegh's *The History of the World* (1614). RB 601553, The Huntington Library, San Marino, CA

through acts of observation (sig. D2v). Ralegh's vision of history, both in his frontispiece and in the body of the text itself, represents a syncretic union of classical, Christian, and empirical understanding that reflects the order of nature as a structure of representation. And textual and visual mediation begin to stand in for the natural world itself in the work

Like Ralegh, though working in a different genre, the English alchemist and mystic Robert Fludd also foregrounds divine nature as a structure of representation in his *Origin and Structure of the Cosmos* (*Utriusque cosmi historia*, 1617). Unlike Ralegh, Fludd based his account on Plato and a tradition of natural magic, offering a quite different understanding of the mathematics of the universe than that of Galileo.[84] Yet, like both Ralegh and Caxton, Fludd approaches the order of nature as a structure of representation that can be studied through print. Fludd's book included an engraving by Johann Theodor de Bry's of "The Mirror of the Whole World of Nature and the Image of Art" (*Integra Naturæ speculum Artisque imago*) (Fig. 2.8).

De Bry's representation of the *ordo creationis* reveals a geocentric universe in which the fixed spheres of the sun and planets surround earth, demonstrating the enduring hold of a pre-Copernican model of the universe well into the seventeenth century. Fludd earned the ire of observational astronomers such as Johannes Kepler, who attacked him in print for his "poetical and oratorical" vision of the cosmos.[85] The two authors engaged in a polemical debate in Latin treatises that spilled from the press over a period of four years, with Kepler accusing Fludd of using an "occult and shadowy manner" (*occultum et tenebrosa*) of writing while also elevating "pictures" (*pictura*) and "pure symbolism" (*meri Symbolismi*) over substance. In return, Fludd criticized Kepler for using an excess of words (*multis verbis*) and for focusing too much on external movement (*motus exteriores*) rather than internal essence (*actos internos et essentiales*).[86] Despite their public argument in print, Kepler and Fludd both also made sharp critiques of Aristotle while expressing faith in the virtues of occult forces. But the differences between the two influential authors also hinged on the status of words and images as tools of knowledge. Kepler and Fludd both emphasized the power of images to convey scientific truths, but, as Robert S. Westman notes, they held radically different ideas about *how* images could represent those truths. Fludd drew on Dürer in arguing for the necessity of knowledge of the created order of nature on the part of painters and illustrators. In a series of woodcuts based on Genesis and in illustrations throughout the *Origins of the Cosmos*, Fludd experiments with

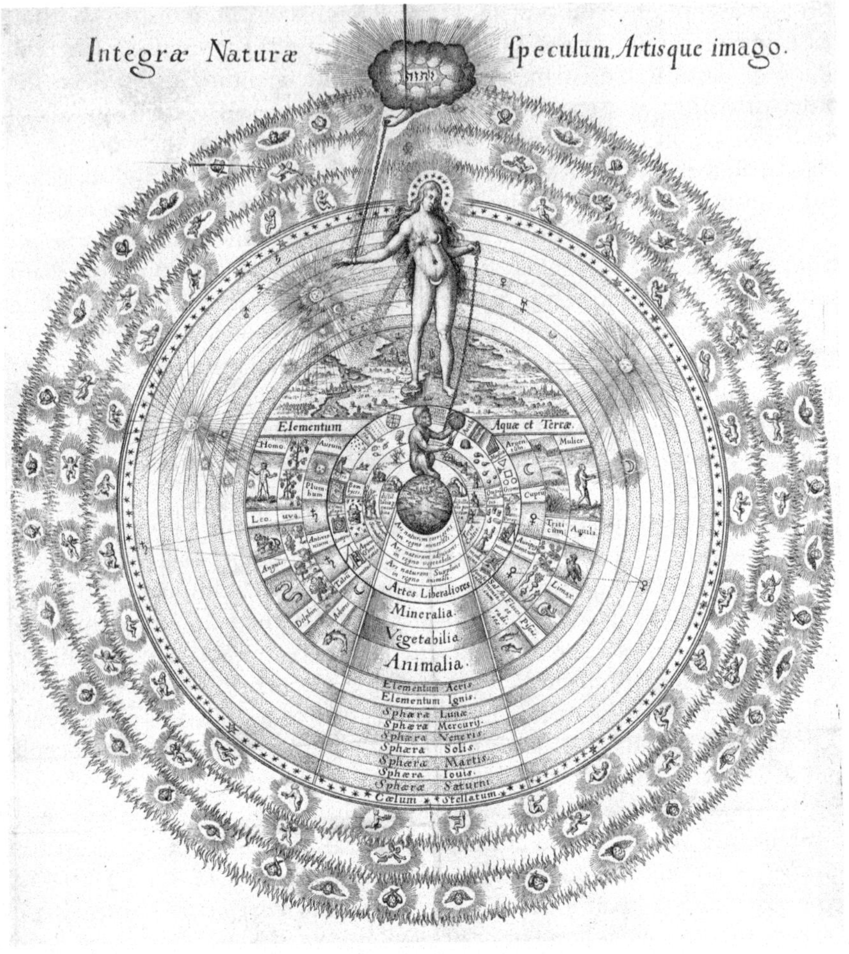

Fig. 2.8 Johann Theodor de Bry's engraving of "The Mirror of All Nature and Image of Art" from Robert Fludd's *Origin and Structure of the Cosmos* (1617–1618)

images that have the capacity to bring into material form aspects of the original creation of the universe. Working in a Christian-hermetic tradition, Fludd believed that both the words and especially the images he created had the capacity to participate in the materialization of original creation.[87]

The engraving by de Bry also pictures the order of nature in and through a series of mimetic relationships. The Tetragrammaton indicating God's formative presence connects to the nude figure of Nature via a chain of being, while Nature herself guides the hand of an ape who holds a compass.[88] This ape symbolizes imitation and enacts the dictates of Nature, who herself follows God. Radiating from Earth and stretching just as far as the ape is tall are four concentric circles representing several "arts." The circle closest to the planet earth depicts "art correcting nature in the mineral realm" (*ars naturam corrigens in regno minerali*[*s*]). The two images illustrating this principle both represent distillation, a process central to alchemy as practiced in Egypt, Iran, and early modern Europe. Moving outward, the next two circles depict "art aiding nature in the vegetable realm" (*ars naturam adjuvans in regno vegetabili*[*s*]), illustrated with visions of farming and horticulture, and "art supplementing nature in the animal realm" (*Ars naturam Supplens in regno animali*[*s*]), which includes silkworms producing silk and the chemical arts associated with medicine. The fourth circle showing the *Artes Liberaliores* (the "more free arts") passes through the head of the ape, signifying the mental agency required for the arts, as well as the orbiting moon that the ape is calculating (see Fig. 2.9).

The center circles of the image depict the disciplines of the medieval *quadrivium* (music, geometry, mathematics, and astronomy) without the verbal arts of the *trivium* (rhetoric, grammar, and logic) normally included in the seven liberal arts. In the place of these missing disciplines, the engraving depicts seven additional arts (moving clockwise from the bottom center): the study of motion, the marking of time, cosmography, Geomantia (an ancient type of divination based on the interpretation of soil, rocks, and sand), the arts of perspective, painting, and architectural design (of military fortifications). This depiction of mechanical arts marginalizes linguistic arts of representation in favor of visual arts such as painting and architecture. Yet it also centers all arts as both the product of nature and the means by which nature can be studied and understood. The concentric circles in this influential "mirror" of nature inscribe the order of nature in both pictorial and textual representations while also aligning its own visual representation with the key to understanding that order itself. Mimetic representation is both the logic of the Great Chain of Being depicted in the image and the divine gift that fosters knowledge about the world. Like other printed books we have examined in this

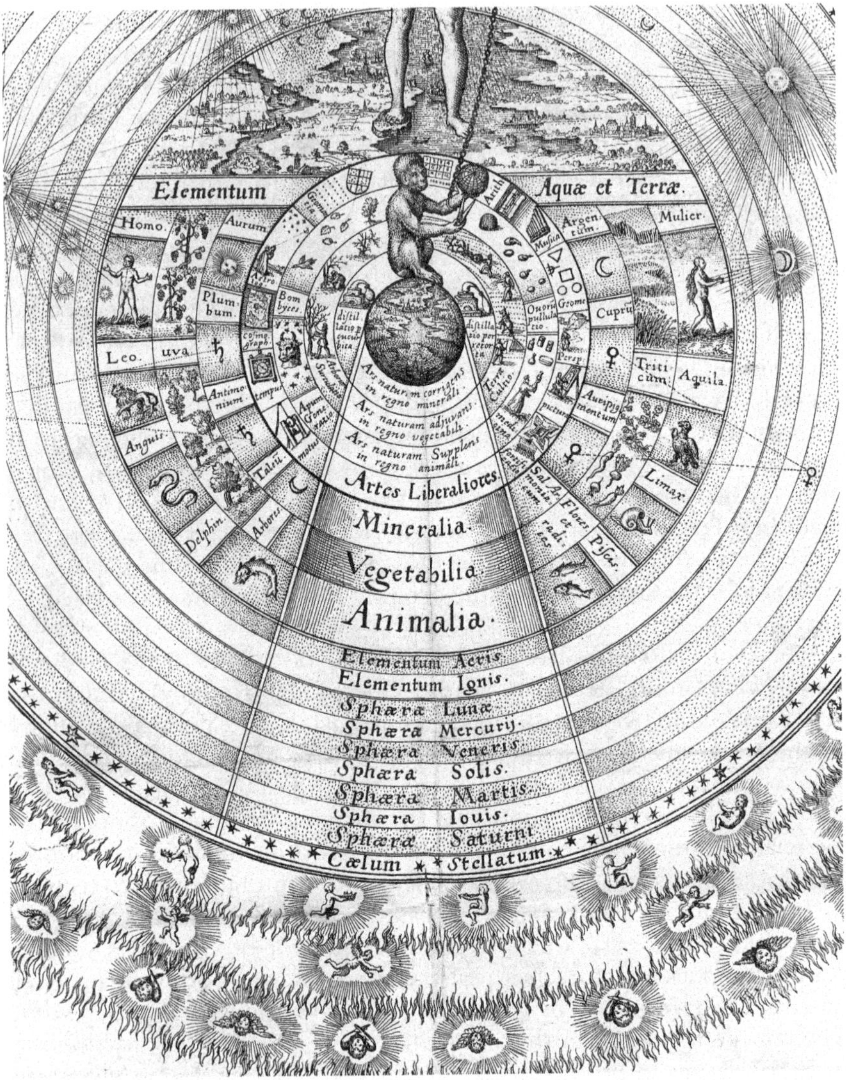

Fig. 2.9 Johann Theodor de Bry's engraving of "The Mirror of All Nature and Image of Art" from Robert Fludd's *Origin and Structure of the Cosmos* (1617–1618) (detail)

chapter, Fludd's order of nature is thus the ground for a radical revisioning of the order of knowledge and the human arts, which, as the image suggests, itself depends on the growing power of print as a mode of visual and textual representation.

Notes

1. Augustine, *Augustine on the Psalms*, 1: 272.
2. Harvey, *De motu cordis*, 8, with English translation in *The Anatomical Exercises*, ed. Keynes, xiii. Robert Boyle also alludes to the "magnificent fabric of the universe" in *Some Considerations Touching the Usefulness of Experimental Natural Philosophy*, 3.
3. Harvey, *Anatomical Exercitations*, sig. a5.
4. The *Oxford English Dictionary* cites the first instance of the meaning of "dissection" as "the methodical cutting up of an animal or a plant, for the purpose of displaying its internal structure" in 1605 (s.v., "dissection," def. 2). The charter of the Barber-Surgeon Company, patronized by King Henry VIII, describes the value of conducting anatomies to "further and better knowledge, instruction, insight, learning, and experience in the said science." See Cunningham, "The Kinds of Anatomy," esp. 11.
5. Thomas Tymme, *Dialogue Philosophical*, sigs. A3-A3v.
6. Jesse M. Gellrich discusses the medieval hermeneutics of the Book of Nature in relation to the bible in *The Idea of the Book in the Middle Ages*, 18 et. passim.
7. Curtius, *European Literature and the Latin Middle Ages*, 310.
8. Ibid., 310–11.
9. Bacon, *The Advancement of Learning*, ed. Kiernan, 30.
10. Robert Boyle, *The Usefulness of Natural Philosophy*, in *Works*, ed. Davis, Cyril, and Hunter, 2:62–3, 19.
11. On Bacon's attitudes toward the language of nature, see James J. Bono, *The Word of God and the Languages of Man*, esp. 207–43.
12. See Katherine Calloway, *Literature and Natural Theology*, 38–9.
13. The manuscript from which these words derive was housed in the personal library of the Earl of Oxford and now resides in the British Library. It is reprinted in *Works*, ed. Spedding and Ellis, 6:34.
14. Bacon, *Novum Organum*, ed. Jardine and Silverthorne, 82. See also William Eamon, *Science and the Secrets of Nature*, 9.
15. Bacon, *The Advancement of Learning*, ed., Kiernan, 109.
16. Bacon, *The Advancement of Learning*, ed. Kiernan, 64–5.

17. On the emergence of a more modern understanding of "experiment" in Bacon and the period more broadly, see Schmitt, "Experience and Experiment." See also Lisa Jardine's discussion in *Francis Bacon: Discovery and the Art of Discourse*, 136–8.
18. The Latin text of this poem is in the *Poetae Latini Medii Aevi 2* of the *Monumenta Germaniae Historica*, ed. Dümmler et al., 186, trans. Irvine, *The Making of a Textual Culture*, 14.
19. Irvine, *The Making of a Textual Culture*, 8.
20. See *The "Etymologies" of Isidore of Seville*, ed. and trans. Barney et al., 231.
21. Ogilvie, "The Many Books of Nature," 30.
22. Galileo, *The Assayer* (1623), trans. Drake in *Discoveries and Opinions*, 237–8.
23. Galileo's letters on sunspots are included in *Discoveries and Opinions,* trans. Drake, 113. For a discussion of how Galileo employed language to bridge Platonic formalism with the sensible world, see Drake, "Galileo's Language."
24. On Galileo's revision of Platonic mathematics and his designation of the cosmos as both infinite and coequal with earthly knowledge, see Alexandre Koyré, "Galileo and Plato." As Margaret Healy notes, both St. Paul and John Dee also reveal the created world as divinely numbered. See *Shakespeare, Alchemy, and the Creative Imagination*, 80.
25. For more on this relationship, see Drake, *Essays on Galileo and the History and Philosophy of Science*, 1:53–54.
26. Galileo, *Sidereus Nuncius,* 35.
27. Galileo, from letters on sunspots, in *Discoveries and Opinions of Galileo,* ed. Drake, 90–91.
28. Marchitello, *The Machine in the Text*, 108.
29. On the printing history of Galileo's study of sunspots, see Drake's introduction to *Discourse on Bodies in Water* (1612), ed. Drake, trans. Salusbury, ix. See also Marchitello, 104.
30. See Nauert, Jr., "Humanists, Scientists, and Pliny," 73.
31. E.W. Gudger has identified at least 89 editions produced in Europe during the sixteenth century, falling to 43 in the following century and 19 in the eighteenth century. Dwindling editions in later centuries indicate his waning influence. See Gudger, "Pliny's *Historia Naturalis*, the Most Popular Natural History Ever Published."
32. Nauert, "Humanists, Scientists, and Pliny," 80.
33. See Eisenstein, *The Printing Revolution*, 64–73. Such a claim does not necessarily contradict the argument of Adrian Johns that printed editions varied widely and contained multiple errors.
34. See Doody, *Pliny's Encyclopedia*, 23.
35. For more on humanist efforts to repackage classical texts in Tudor England, see Nicholas Orne, *Education and Society in Medieval and Renaissance England*, 29.

36. Doody, 23.
37. Pliny, *A Summary of the Antiquities and Wonders of the World*, sig. A2.
38. Pliny, *The History of the World*, trans. Holland. Both the "Preface to the Reader" and the Letter from "H.P." from which these quotations derive were inserted without page or signature number before the table of contents.
39. Ibid., from unpaginated "Preface to the Reader."
40. Editors often note that Othello's vow to follow the "icy current and compulsive course" of the Pontic see, that "ne'er keeps retiring ebb, but keeps due on / To the Propontic and the Hellespont" (3.3.446–50) quotes directly from Holland's translation, while T.W. Baldwin suggests that Shakespeare may also have drawn on Latin editions of Pliny that circulated throughout England. See his "A Note upon William Shakespeare's Use of Pliny."
41. Roland Frye remarks about this passage that "it is difficult to find a more succinct, accurate, and memorable summary of the Christian theology of nature." See *Shakespeare and Christian Doctrine*, 219.
42. Rebecca Bushnell highlights the "unstable" duality of good and bad in plants as well as their relation to human virtue in her survey of "Vegetable Virtues" in her contribution to *Shakespeare's Botanical Imagination*, esp. 54–5.
43. Jacques Derrida traces both the literal and figurative sense of the *pharmakon* in a number of Plato's dialogues in his section on Plato's pharmacy in *Dissemination*, 61–171. For a reading of this passage in light of Derrida, see Girard, "Lévi-Strauss, Frye, Derrida and Shakespearean Criticism," 58.
44. Finsbury fields were just west of the Theater, which was a short walk north of the city wall of London by way of Moorgate. According to Stephenson, such fields "abounded in wild flowers." See *Shakespeare's London*, 292.
45. Agrippa, *Opera*. Cited and translated by Brian Copenhaver in his chapter on "Magic" in Park and Daston, eds., 519. Although the printed editions of Agrippa did not appear in London until 1650, his ideas were interpolated by many others and find their way into other works, including the Antiphoner Notebook and the Boxgrove Manual. See, e.g., *Making Magic in Elizabethan England*, ed. Klaassen.
46. Martin Luther, *Last Will and Last Confession*, sig. c4.
47. William Tyndale (trans.), *The New Testament*, sig. C5v.
48. William Turner, *A New Herbal*, sigs. A2-A2v.
49. Nicolas Culpepper, *The English Physician*, sig. B1.
50. For an account of the early modern renaissance in herbals, see Agnes Arber, *Herbals, Their Origin and Evolution*. On Brunfels and the new emphasis on drawing from direct observation, see Brent Elliott, "The World of the Renaissance Herbal."

51. We do not know if Gerard ever saw a play by Shakespeare, but he does say in his *New Herbal* that he attended plays at The Theater, 950. On the possible acquaintance and even friendship between the two men, see Margaret Willes, *A Shakespeare Botanical*, 16.
52. The English botanist and editor Mark Griffiths speculates that this figure in the 1597 version of the title page was an image of Shakespeare. The claim circulated widely in popular media but was dismissed by many Shakespeare scholars, including Stanley Wells. For a discussion of the attribution and reactions in relation to plants in Shakespeare, see Susan C. Staub's introduction to *Shakespeare's Botanical Imagination*, esp. 11–14.
53. John Gerard, *Herbal, or General History of Plants*, rev. ed., sig. ¶4.
54. Foucault, *The Order of Things*, 35.
55. Ibid., 34.
56. For a critique of Foucault, see Ian Maclean, "Foucault's Renaissance Episteme Reassessed." See also Crane, *Losing Touch with Nature*, 33. Both authors nevertheless acknowledge Foucault's enduring influence.
57. For a discussion of this ontological problem in Exodus, see Noam Reisner, *Milton and the Ineffable*, 16–20.
58. Derrida's theory of logocentrism in *Of Grammatology* overlooks this duality, in part because he ignores religious ideas in his account. His primary sources are secular authors such as Rousseau, Pierce, Saussure, and Lévi-Strauss.
59. See, e.g., Robert Lafont, *Anthropologie de l'écriture*, 115.
60. Origen, *In Leviticum* 1.1. See the *Patrologia cursus completes*, 12:405.
61. Jerome, *In Psalmum* 147. Cited and translated by Jared Ortiz in *Deification in the Latin Patristic Tradition*, 27.
62. The author known as Pseudo-Dionysius probably dates to the late fifth and early sixth century C.E., but his writing is deliberately designed to imitate the style of the earlier figure of St. Dionysius the Areopagite, a member of the Athenian judicial council (known as "the Areopagus") in the first century C.E. and the first convert to Christianity by St. Paul. See Sarah Coakley's introduction to *Re-thinking Dionysius the Areopagite*, 1–10.
63. Denys, *The Divine Names* (VII:3). Translated by Karen Armstrong in *A History of God*, 127.
64. The editor is Philip Schaff. See his introduction to *Augustine*, 28.
65. See, e.g., Augustine, *De trinitate* (4.xxi.30). For a discussion, see Eugene Vance, "Saint Augustine: Language as Temporality," in *Mimesis* ed. Lyons and Nichols, 20–25.
66. And he was not alone. For a study of a longer tradition, see William A. Dyrness, *Reformed Theology and Visual Culture*, 26.
67. Augustine works through the different layers of spiritual, angelic, human, and literal language in book I of *The Literal Meaning of Genesis*, 21–54 et. passim.

68. For more on these analogies, see Bernard Montagnes, *The Doctrine of the Analogy of Being According to Thomas Aquinas.*
69. Translated by Karl Young in "The *Speculum majus* of Vincent of Beauvais," 6.
70. Ritamary Bradley, "Backgrounds of the Title *Speculum* in Medieval Literature," 103, 110. See also Mary Franklin-Brown, *Reading the World*, 272.
71. Hugh of Saint-Victor, *Expositio in Regulam S. Augustini* 12, PL 176, col. 924, trans. Mary Franklin-Brown, *Reading the World*, 272. The Latin reads "nobis...nostrum interiorem repraesentat imaginem; ostendit quid formosum, quid deforme sit in anima, et qualiter pulchritudo justitiae debeat observari, qualiter debeat virtutum decor componi; quomodo vitiorum macula debeat abstergi."
72. Lacan speaks of the "illusion of autonomy" of the ego. See "The Mirror Stage," in *Écrits*, 6.
73. As Sargent remarks in his modern edition of the work, only the Wycliffite translation of the bible, *The Canterbury Tales*, and a moral poem called *The Prick of Conscience* can be found in more manuscript copies than the *Mirror of the Blessed Life of Christ* before the Reformation. See Nicolas Love, *The Mirror of the Blessed Life of Jesus Christ*, ix.
74. For an account of this reading experience in relation to Love's text, see Laurelle LeVert, "Crucifye hem, Crucifye hem."
75. See Sargent's introduction to Love's *Mirror*, xvii, which draws on Kantik Ghosh, *The Wycliffite Heresy.*
76. For an account of European print history that incorporates knowledge of printing elsewhere in the world, see *The Book Worlds of East Asia and Europe.*
77. Daston, "Observing Nature," 127.
78. William B. Ashworth, Jr., "Emblematic Natural History of the Renaissance," 23–24.
79. William Caxton, *Here Beginning the Book Called the Mirror of the World*, sig. a4.
80. Caxton, *Mirror of the World*, sig. a4.
81. Wolgemut's studio produced the 645 woodcuts from which Koberger produced more than 1800 illustrations.
82. Translation by Jeffrey Chipps Smith in *Nuremberg, A Renaissance City*, 94.
83. Walter Ralegh, *The History of the World*, 2. All quotations from this text are to this first edition, hereafter cited parenthetically by page number. For a modern edition, see Ralegh, *The History of the World*, ed. Patrides, 85–6.
84. On Fludd's relation to mathematics, see Allen G. Debus, *Man and Nature in the Renaissance*, 12.
85. Cited by Patrick Boner in *Kepler's Cosmological Synthesis*, 8. Boner also acknowledges the many points in common between Kepler and Fludd throughout his book.

86. These passages are quoted and translated in an essay on the status of words and images in the larger dispute between Kepler and Fludd by Robert S. Westman in "Nature, Art, and Psyche: Jung, Pauli, and the Kepler-Fludd Polemic," in *Occult and Scientific Mentalities*, ed. Vickers, 180.
87. Ibid., 203–6.
88. Principe reads the image as an expression of the confluence of nature, art, and God but interestingly overlooks the inclusion of painting and perspective in the *Artes Liberaliores*. See *The Secrets of Alchemy*, 195–8.

CHAPTER 3

Staging Nature

The printed works analyzed in the previous chapter redeployed old tropes in new ways to recenter the natural world as a site of scientific and theological inquiry. This technological reframing of the order of nature elevated new modes of representation as central to early modern scientific inquiries in fields such as zoology, astronomy, and botany. Such books and the logic of *grammatica* that they employed also confirm that early modern efforts to return to nature were not triumphs of unmediated observation of the *ordo creationis*. Rather, as we have seen, they mixed older and newer epistemologies and textualities with technologically enhanced forms of mimesis.

These efforts to return to nature were also embedded in complex questions of theology. Peter Harrison has suggested, for example, that Protestant antipathy toward allegorical and symbolic hermeneutics of earlier centuries inspired greater emphasis on the literal interpretation of the Book of Nature.[1] Although Harrison's thesis has been criticized by some, his larger point is that new efforts to discern the legibility of nature shifted early modern relations between words, images, and objects as modes of representation.[2] Grounded in narratives of divine creation, the early modern return to nature also depended on new modes of visuality that were sometimes, as we saw in the case of Galileo

A. Kitch, *Shakespeare's Theater of Nature*, Palgrave Studies in Literature, Science and Medicine,
https://doi.org/10.1007/978-3-031-78082-0_3

and Holland, overtly theatrical. The remainder of this book will focus on forms of theater that engaged debates in theology and natural philosophy to test the legibility of nature. This chapter argues that a major shift in modes of theatrical representation occurred in the early sixteenth century with reference to the order of nature. Early Tudor interludes by Henry Medwall, John Rastell, and John Heywood reconfigured divine nature as a subject of dramatic representation, drawing on previous traditions of dramatic mimesis and informed by nascent printing practices in England. Plays by these and other authors followed several centuries of innovations within liturgical drama and cycle plays that responded to efforts within the Catholic church to assimilate Aristotelian natural philosophy and to evaluate the relation between divinity and representation more broadly.

By way of introducing these innovations, I first survey modes of liturgical drama beginning with the *quem quaeritis* trope and the theological questions about representation that they foreground. In extending my analysis to several later modes of dramatic representation, I show how struggles to situate modes of drama within questions of divine legibility shaped a wider range of drama in late medieval England. Following readings of Tudor interludes by Medwall and others that were written and performed in the first half of the sixteenth century, I conclude the chapter with interpretations of Christopher Marlowe's *Dr. Faustus* and Shakespeare's *Romeo and Juliet* that demonstrate the enduring influence of such problems of knowledge and visibility in new forms of professional theater in Elizabethan England.

Older accounts of the transition from liturgical tropes to cycle plays to morality plays by E.K. Chambers, Karl Young, and others have been rightly criticized for employing a Darwinian model of "evolution" that posits a linear history of theatrical mimesis stretching from liturgical ceremonies to Shakespeare. For Chambers, for example, plain chant of the tenth-century church service evolves historically to form dialogue in versions of the *quem quaeritis* trope-turned-Easter performance.[3] Drawing on a larger archive of records from the *Records of Early English Drama* (REED), more recent scholars have challenged these assumptions and offered different frameworks for understanding representation as a set of practices focused on questions about the incarnation of Christ and the problems of representation posed by such bodily transformation.[4] Sacraments such as baptism and the eucharist, for example, combined matter and form in unique ways, as Thomas Aquinas outlined in his *Summa Theologica*.[5] Michael Kobialka has argued that questions about Christ's body informed new social, cultural, religious, and scientific debates that shaped a proliferation of new modes of drama in medieval England.[6] Like Kobialka, I resist arguments that describe the movement of ludic activity from the church to the street to the playhouse as process of secularization. My account emphasizes Christian appropriations of Aristotle and questions of natural philosophy focused on the representation of divinity as significant influences on genres and forms of dramatic mimesis in medieval and early modern England.

The emergence of new modes of dramatic mimesis within the Catholic church beginning in the twelfth century coincided with radical new questions about Christian theology that responded to the rediscovery of classical natural philosophy by way of Latin translations of Islamic copies of classical Greek texts. Earlier Christian theologians, especially Tertullian, had dismissed Aristotle and other Greek philosophers as anti-Christian, in part because of Aristotle's seeming rejection of divine providence and the immortality of the soul. Following Augustine, Christian theology took on a more Platonic orientation, but beginning in the early twelfth century the church relaxed its ban on Aristotle, fostering Aristotelian works of theology by Albertus Magnus and Thomas Aquinas in particular. As David C. Lindberg remarks about this turbulent time in the history of the church, "Aristotelian philosophy and Christian theology became allies [...] and in the long run Aristotle became the official philosopher of the Catholic church."[7]

Representing God

The organized study of theology at the newly established universities of Paris and Oxford also provided important sites for elevating philosophical and theological questions in medieval Europe. A perennial problem concerned how to understand and represent Christ's divine and human nature simultaneously, especially since the divine form was invisible. In his *Sententiae* (1155–57), for example, the Bishop of Paris Peter Lombard drew on Aristotle to question how to worship Christ's human and divine natures simultaneously. Aquinas followed Lombard in his study of a similar problem of representation, helping to devise a wider continuum of material divinity and the appropriate modes of worship for sacred objects. Aquinas reserves *latria* ("adoration") for God alone, *dulia* ("simple veneration") for divine objects that are beautifully created but should not be worshipped, and *hyperdulia* ("extreme veneration") for the Virgin

Mary.[8] As Michael Camille notes, the history of art in this period of European history was largely devoted to the question of how to transform visible objects relegated to a secondary status in relation to an invisible divinity into "meaningful spectacle."[9]

Such questions found their way into the influential Fourth Lateran Council of 1215, which was dedicated in part to natural philosophical questions such as the problem of the "unity of nature" with regard to the Holy Trinity, a problem about how divine essence can be distributed equally among the Father, Son, and the Holy Spirit.[10] In asserting a universal unity in both essence and form of the Holy Trinity, the Lateran Council drew attention to the extended material presence of God in a way that justified greater observation of the Book of Nature. The council also recommended practices that helped to stimulate new modes of mimesis within the church. At stake in both theological debates over performance and the drama shaped by such debates were questions about divine representation. The Fourth Lateran Council also encouraged an expansion of the custom of elevating the Host as a form of "meaningful spectacle." This elevation, which involved priests holding the communion wafer and goblet of wine aloft before the assembly following each ritualistic taking of communion, began in Paris and spread rapidly throughout Europe.[11] Often accompanied by music, the elevation of the host defined a new kind of visual sacramentalism that was controversial since some preferred communion through physical eating (*per gustum*) rather than through visual spectacle (*per visum*). As Kobialka notes, a new emphasis on bodily presence in the eucharist (*Hoc est corpus*) was part of a larger strategy of "producing and disseminating the visibility of the sacramental body" in the period.[12] This ritualistic elevation of the Host also spurred new modes of liturgical drama.[13]

An early text of such liturgical drama responded to such questions in new ways. Enacted during the Easter liturgy, the *quem quaeritis* trope centers on the narrative of Christ's death and resurrection and the ensuing visit by the three Marys to the tomb of Christ. The simple question posed by the three women in the trope—*quem quaeritis* or "Whom do you seek?"—receives the familiar but frustrating answer that reproduces the famous passage from Exodus in which God both names and conceals himself: "Non est hic"—"he is not here."[14] Following a similar logic, the *quem quaeritis* represents Christ's resurrection without staging it directly. The liturgical trope was a mode of visual representation that responded to the problem of how to embody and visualize divinity. The absent Christ is reincarnated through words, chant, and embodied action during the Easter liturgy. In the process, he is redistributed throughout the ceremonial space of performance—in the processional cross, the cross in the nave, and also in the eucharist, as well as in the dispersal of Christ's love to parishioners. In some versions of the trope, including the early Winchester text, the ceremonial performance draws attention to the empty space where Christ's body had been, leading to the announcement of his resurrection (*Alleluia, resurrexit Dominus*). Other versions of the trope employed a shroud covering the cross that was removed to show the absence of Christ, and in others this shroud promised Christ's return, following the implicit belief that he would appear when the shroud itself did. Later instances of the *quem quaeritis* in Prague and elsewhere emphasize the visibility of the eucharist as the body of Christ in relation to the invisibility of Christ as risen from his tomb. The doctrine of Resurrection governs the logic of such displays, as in the related trope of the raising of Lazarus, which prefigures the Resurrection of Christ in the New Testament. The miracle of resurrection is a supernatural event that cannot be accounted for by the laws of nature. It transcends nature and is therefore outside of the scope of the *quem quaeritis* trope's mimetic power.

The rising popularity of the *quem quaeritis* trope joined renewed interest in the ancient work of *The Celestial Hierarchy*, originally attributed to Dionysus, which was translated by Hugh of St. Victor and Robert Grosseteste in the thirteenth century. *The Celestial Hierarchy* focused on the ways in which sacred texts "visualize" invisible and divine forces, locating the created, material world as a necessary intermediary between profane witnesses and the sacred truths they seek.[15] The emphasis in the text on the visible body of Christ as the material manifestation of spirit had implications for understanding the Christian liturgy as a mode of dramatic *agon*. Honorius of Autun (d. c. 1140), for example, identified a conflict between commemoration and the doctrine of Incarnation, creating a dramatic conflict around which some works of medieval drama centered.[16]

Drama performed outside of the church in cycle plays such as those produced in York beginning in the fourteenth century attempted to stage God more directly, sometimes even opening with a God figure who proclaims his status as original creator of nature. But this new mode of drama remained tainted by anxieties of blasphemous imitation, following biblical examples such as the vexed attempts by Adam and Eve to imitate God by eating the fruit of the Tree of Knowledge. The appearance of the story of the Fall of Lucifer in several extant Corpus Christi plays foregrounds such dangers of mimesis. Consider, for example, the Towneley *Creation and Fall of the Angels,* which opens with the figure of God, Deus, enthroned in heaven and surrounded by angels. In his lengthy monologue, Deus claims to be "without beginning" and to contain "all manner" of things in his thoughts (lns. 7, 13).[17] He is the alpha and omega, the "first" and the "last" (lns. 1–2), a creator of earth and oceans and of the stars and seven planets. He is both contained and dispersed across the "three" persons of the Holy Trinity, and his divine presence cannot be captured by language alone (ln. 10). He takes responsibility for the creation of the divine order of nature, but he does not make that order open or accessible to forms of knowledge. Following apophatic logic, he exceeds the forms of representation used to portray him, just as the vast darkness of the divinely created cosmos can never be known completely.

By contrast, the figure of Lucifer, who appears to fill the void created by the withdrawal of Deus from his throne, is more present and accessible to the audience. Lucifer describes himself as "so fair and bright" (ln. 82), as a participant in the "light" (ln. 83) of divine creation who is "a thousandfold / Brighter" than Christ (lns. 88–9). His essence is purely material and fully dependent on visual representation.[18] Unlike Deus, he does not transcend the world and has no relationship with invisible forces of the universe. He does not draw attention to the order of creation but, rather, to himself in all of his visual splendor. Lucifer thus exemplifies the challenge of theatrical representation in medieval England and, in a later staging, in early modern England. His brilliant theatrical visibility also reminds spectators of the limits of theatrical representation, which the Towneley *Creation* aligns with diabolic sin.

The twelfth-century Anglo-Norman *Play of Adam* is another innovative experiment in dramatic mimesis that situates the problem of representing divinity in relation to the order of nature. The Latin title of the work, which exists in only one copy in Tours and was probably produced in parts of England where French was widely spoken, is *Ordo Repræsentationis Adae*, or "Service for Representing Adam." As Eric Auerbach observes in his reading of the play, the drama mingles secular and profane registers in innovative ways, presenting sublime concepts of divinity in vernacular dialects in order to bring biblical knowledge to a broader audience.[19] The *Ordo* depicts the temptation and fall of Adam and Eve, the story of Cain and Abel, and a series of speeches by Old Testament prophets foretelling the coming of Christ. Like Milton in a later century, the author of the *Play of Adam* expands on the sparse details of Adam and

Eve in Genesis to create a psychologically complex story about the Devil's strategies to tempt Eve into disobedience of God after he has failed to succeed with Adam. The work begins with the "Figure" (*Figura*) of God teaching Adam the "lesson" (*lectio*) of his creation of both the heavens and the earth.[20] This Figure reminds Adam that He has "formed" him Out of the "loam of the earth" (lns. 2–3) and provided him with a "worthy companion" in Eve (ln. 9). After Adam and Eve both swear obedience to God and to one another, the Figure points to the garden of "paradise" (*paradisum*) that he personally "planted" and "laid out" (lns. 83). This Eden is a space of eternal life, free from death, anger, sin, and shame (lns. 89–100).

As the Figure of God departs, the paradise of Eden becomes a site of His divine presence and the means of signifying human obedience to Him, as He orders Adam and Eve not to eat the fruit of the forbidden tree (lns. 101–2). Over the course of the drama, the divine garden operates as the medium through which Adam and Eve define their relationship to one another and through which the Devil effects their disobedience to God by tempting Eve to eat the forbidden fruit. When the Figure returns to condemn Adam for his transgression, it is the order of creation itself that registers the consequences of their sinfulness:

> The earth will be cursed.
> Where you will wish to sow your grain.
> It will fail to bear fruit,
> It is cursed beneath your hand;
> You will cultivate it in vain.
> It will deny its fruit to you;
> It will yield you thorns and thistles.
> It will change whatever you sow;
> It will be cursed, as punishment to you. (lns. 426–34)

This corrupted earth following the transgression of Adam and Eve links the first episode of the *Ordo Repræsentationis* to the second, in which Abel offers the "best and fairest" lamb to God as an offering (ln. 644), while Cain offers only a handful of wheat (ln. 649).[21] After the Figure of God accepts Abel's generous offering and disdains the offering of Cain, Cain kills Abel out of spite (lns. 720–23). According to the biblical story, the blood of the murdered brother mingles with this fallen earth, calling out to God for assistance (Genesis 4:10). God curses Cain "from the earth, which hath opened her mouth to receive thy brother's blood from thy hand" (Genesis 4:11). Where the *quem quaeritis* trope follows the logic of resurrection in creating its logic of visibility and invisibility, the *Play of Adam* figures divinity in terms of the *ordo creationis* and the typological anticipation of the birth of Christ. The *Ordo Repræsentationis* thus projects the order of nature as a space of figuration that incorporates both human and divine characters. The fallen earth that registers the sinfulness of Adam and Eve and that produces the meager offering of Cain becomes the medium through which the Christ of the New Testament can achieve human salvation. Like the previous plays considered in this section, the *Play of Adam* stages problems of divine presence by developing new forms of theatrical mimesis that must navigate challenging problems of theology.

Tudor Plays of Nature

Humanist interludes of the newly established Tudor court in sixteenth-century England generated new modes of dramatic mimesis while also foregrounding nature in new ways. The introduction of print to England beginning in the late fifteenth century shaped these theatrical innovations. The author of the earliest known vernacular drama in England was Henry Medwall, a chaplain to the Archbishop of Canterbury, John Morton. His interlude of *Fulgens and Lucrece* (c. 1490s) was a dramatic version of Bunaccorso da Montemagno's humanist treatise, *Of True Nobility* (*De vera nobilitate*), which was translated into English in 1460 and printed by William Caxton in 1481. Dramatizing a tale of two men from different social classes competing for the hand of Lucrece, the daughter of Fulgens, the play foregrounds questions of merit vs. inheritance that were central to the new understanding of courtly virtue based on inherent skill rather than birth or social connections at the newly established Tudor court.[22] The dramatic interlude features two servants, "A" and "B," who comment on the action in ways that imitate the clever servants of Terence and who provide an innovative sub plot based on forms of rhetorical argument derived from Cicero.[23]

Medwall's other interlude of *Nature* (c. 1490s) appears more backward looking, given its use of allegorical characters and plot in the style of popular morality plays. But the interlude is also innovative in the fact that Medwall replaces God with "Nature." As a minister of God, this character of Nature assumes the task of creating "each creature" and of engendering a "spirit of life" in all things, including making the moon wax and wane and the seas ebb and flow.[24] Rather than directing the audience toward the order of nature, however, Nature urges viewers and readers to consult Aristotle, her "philosopher elect" (45). Medwall's Mankind figure at first pays humble homage to such nature by recognizing a "common being" with plants, herbs, and animals who share the same "virtue" (47), but things change as Sensuality and Reason arrive to struggle over his fate. Sensuality asserts her role as the "chief perfection of his nature" (48) and Reason anticipates that the frail body of Mankind is "disposed for to fall" (54), following the example of Adam and Eve. The play then follows a familiar pattern of sin and redemption in staging Mankind's irrational embrace and companionship with Pride and other vices, including Worldly Affection and Bodily Lust, before advanced age and fragility drive him back into the fold.

Despite its allegorical characters and morality-play structure, then, *Nature* represents a shift in dramatic mimesis in relation to earlier forms of European drama by grounding its theological message more fully in the order of nature. Mankind naturally returns to Reason by virtue of the ageing process, which makes his appetite for Sensuality decrease. At stake in the play, as Liza Blake remarks, is the struggle to understand the operation of the order of nature itself: "Medwall finds in theater the perfect way of investigating the role that allegorical drama can—and must—play in understanding not only how our ideas about nature are intimately connected to ideas such as causality, responsibility, and the place of the human within the natural world, but also how staged allegory might be used to engage in debates about nature."[25] Any argument that *Fulgens and Lucrece* looks forward toward later and more "secular" drama while *Nature* looks backward toward religiously oriented morality plays thus ignores the latter's innovative elements. *Nature* adapts a morality play structure to entirely new content in order to address familiar problems about how to stage divinity and the divine order of nature. Both plays were probably designed for performance at Lambeth Palace before Cardinal Morton.[26] And, as Richard Sutton observes, "Nature's opening speech is perhaps the first speech in English drama to give some account of natural science, and in this sense it must have been novel and 'modern' to its first hearers."[27]

Medwall's *Fulgens and Lucrece* also has the distinction of being the first drama printed in England. The lawyer John Rastell, who had studied at Oxford and the Inns of Court, opened his printing shop in London around 1512. The interlude was published around 1517 and Medwall's later play *Nature* was published by Rastell in 1520. Rastell's visually striking printer's device itself foregrounds questions about science, religion, and mimesis that also define the plays he wrote and printed by incorporating Aristotelian natural philosophy and a divinized order of nature (Fig. 3.1). Flanked by the English Royal Coat of arms and the three feathers signifying the Prince of Wales, the woodblock image features a scene of creation ("fiat"), of the Word made flesh. The half-human, half-reptilian figures may be mythological, but their juxtaposition with the four elements also suggests that they represent Adam and Eve. The four elements themselves appear in ascending order from earth to water to air to fire, surrounded by the sun, moon, and stars of the cosmos. The fringed cloth with Rastell's initials on it is being torn by Adam, perhaps representing the veil of the Temple that was supposedly defaced after the death of Christ (Luke 23:45).[28] Taken together, these elements represent a microcosm of the *ordo creationis.*

Rastell was not only a printer but also a prolific playwright who composed several of his own interludes for courtly audiences during the reign of King Henry VIII. In his *Four Elements* (c. 1518), for example, Rastell follows Medwall's *Nature* by structuring his morality play around forms of natural philosophy. The Messenger opens the play with an invocation to the "grace of the power divine" and promises to produce knowledge about "philosophy natural," arguing that the English "tongue is now sufficient" to convey such important matter.[29] The Messenger acknowledges the difficulty of knowing God except through knowledge of his "creatures," which the play will make manifest (lns. 92–4). The interlude promises to make the invisible and ineffable works of God present and visual for the audience. Praising the virtues of education in the sciences, for example, the Messenger acknowledges that the four elements as "daily appear here at eye" (ln. 125) as created by God, noting that the interlude will bring its "matter" before the "presence" of the audience by means of the physical "presence" of the players themselves (ln. 144). He exits to make way for the first such player, who enacts the role of "Natura Naturans" (i.e., creating Nature), the "immediate minister for the preservation" of things on earth (ln. 156). Natura Naturans also urges the audience to learn of higher things, such as the nature of divine creation, through the study of lower things, in order to access the invisible essence of God through observing the representation of the material world on stage.

Following the departure of Natura Naturans, Humanity declares himself the "child and formed instrument" of Nature (ln. 219). Nature begins the metaphysics lesson by summarizing the four elements and then

Fig. 3.1 John Rastell's printer's device

encourages Humanity to use his own "reason, wit and understanding" in order to "Know thy maker and cause of thine own being" (ln. 298) as well as to "know and laud the high God etern[al]" (ln. 302). In other words, Humanity must learn by means of the order of nature. To aid in this endeavor, Studious Desire appears, turning Humanity's attention to the earth and the process of Aristotelian "generation" in herbs, plants, stones, and metals (lns. 330, 332). But first, Humanity confesses that his reason is inadequate to guide him toward knowledge about things that cannot be seen, such as the shape of the earth and its position in the cosmos. Studious Desire draws his attention to the empirical details that he can know, such as the appearance of the stars and moon from night to night, which indicate that the earth is in the middle of the "firmament" (ln. 361).

With the abrupt entry of Sensual Appetite, who urges Humanity to merriment at the tavern using the popular verse form of tetrameters, the figure of Experience appears to answer questions about geography, such as the number of miles between London and Jerusalem (lns. 678–83). He also points to England's place on a giant globe of the earth, a prop composed of printed woodblocks produced by Rastell's own printing house. Once Humanity leaves his sensual escapades long enough to return to the company of Studious Desire and Experience, he accepts the roundness of the earth based on the appeal to common experience of such phenomena as the appearance of the mast of a ship at a long distance on the open ocean. Even though several pages of the only known copy of the play are missing near the end, it is clear that Humanity falls back into his relationship with Sensual Appetite and a new character named Ignorance, so that Natura Naturans must remind him on his return at the end of the interlude that even though sensual delight must be satisfied, it will not bring happiness or "holy delight" in the same way that the study of nature will.

In mixing sacred and secular ideas, Rastell's *Four Elements* also draws on modes of natural philosophy to question divine creation. In its mingling of music, allegory, and representations of everyday London life (e.g., in the tavern scene), the play unifies the "presence" of a divine God in and through nature with the "presence" of the players themselves who enact the story. It self-consciously incorporates print technology to make the order of nature both visible and "present" for readers and audience members, especially by utilizing a large printed globe produced by Rastell as a central prop. Just as the four elements combine invisibly but have visible consequences, the space of the stage for Rastell embodies invisible features of the order of nature. The interlude also engages in formal experimentation as it attempts to redefine the role and scope of the theater as a medium for staging problems of theology and natural philosophy in new ways.

Medwall's *Nature* and Rastell's *Four Elements* influenced several later works of drama in early Tudor England, including *Wit and Science* (c. 1535) by John Redford, a chorister of St. Paul's.[30] Redford's interlude stages the wedding of Wit, a son of Nature, and Science, the daughter of Reason. Conflict emerges initially in the play between the "glass" or mirror of Reason and the "goodly picture" of Wit that has been sent ahead to Reason to woo her as "his own image sure" (138). Before he can appear before her, however, Wit must first overcome his own "pride" in his accomplishments (141). Predictably, Wit ignores Instruction and falls in with the vice figures of Idleness and Ignorance instead. When he finally does appear before Science, he wears the garb of Ignorance. In response, Science and Experience leave him to ponder his own reflection in the glass of Reason, which allows him to recognize his own fallen condition. Wit laments that, instead of all of the glories of the World that she might have brought him, including Favor, Riches, Worship, and Fame, he instead has won only Hatred, Beggary, and Open Shame. Scourged by Shame, Wit accompanies Reason to receive education from Instruction, Study, Wit, and Diligence, who leads him to Mount Parnassus by overcoming Tediousness. Wit's chivalric act of defeating Tediousness wins the heart of Science at last, though both she and Experience warn Wit that he must still "use" Science in a noble fashion to honor God (173).

The character of Science embodies the kind of learning that Redford's own pupils might have acquired through his instruction in forms of natural philosophy. The play both invokes and visualizes the mirror of Reason as an instrument of insight and character formation over and against the flattering portrait of Wit that Confidence sends ahead to Science in order to seduce her. Like the earlier Tudor interludes considered above, the overtly religious content of *Wit and Science* has been replaced by divine natural philosophy, as Redford describes Science as God's "gift" and Wit vows to use this gift in the service of God's "honor" (174). When Science rejects the praise of Fame, Worship, Favor, and Riches, she is rejecting the values associated with "the World" (159). But, of course, the assembled actors together with their costumes and the very setting of the play are also part of such a world.

This world is also central to the play's pedagogical function. Wit must learn to appreciate Science for the ways in which she represents God, which is to say for the ways in which she represents what cannot be represented. At one point, Shame enters with a whip, while Reason appears to chastise Wit for his broken promises. As Shame whips Wit, the latter kneels before Reason, who offers forgiveness if Wit vows to follow Instruction and his companions, Study and Diligence. Wit joins with Instruction and his friends to defeat the monster Tediousness, winning the confidence and support of Science and her parents. In the climactic song, the relation between seeing, knowledge, and the mirror of theater becomes explicit:

> WIT AND HIS COMPANY
> O lady dear,
> Be ye so near
> To be known?
> My heart you cheer
> Your voice to hear;
> Welcome, mine own!
> SCIENCE AND HER COMPANY
> As ye rejoice
> To hear my voice
> From me thus blown,
> So in my choice
> I show my voice
> To be your own.
> WIT AND HIS COMPANY
> Then draw we near
> To see and hear
> My love long grown!
> Where is my dear?
> Here I appear
> To see mine own.
> SCIENCE AND HER COMPANY
> To see and try
> Your love truly
> Till death be flown,
> Lo! here am I
> That ye may spy
> I am your own. (ll.171–95)

The dialogue suggests that we can only know Scientia through her body and voice, which she reveals to Wit by means of theatrical performance. The marriage of Wit and Science also reaffirms the union of the bride's parents, Reason and Experience, whose marriage signifies a victory not just for the allegorical characters involved, but also for the merging of natural knowledge and embodied performance that the play enacts. The mirror at the heart of Redford's humanist drama thus provides not just a passive reflection but also generates active and transformative knowledge that reforms Wit. Drama operates here as a test of identity and a means of revealing the true self underneath a false appearance. The fact that Wit holds the mirror up to the audience suggests that the audience itself is part of this transformative process.

Like Rastell, the court musician John Heywood was also a humanist at the court of Henry VIII who engaged in both playwriting and printing. His works for the stage consistently foreground the order of nature as a source of representation. Heywood, who was married to William Rastell's sister Elizabeth, was probably a member of Henry VIII's court between 1515 and 1528. His interlude of *The Play of the Weather* (1533) begins in the "presence" of the god Jupiter, who presides over a parliament of gods and goddesses assembled to redress conflicting demands of human petitioners.[31] Jupiter hears the various demands of suitors representing different segments of the social world—the gentleman desires good weather for hunting, for example, while the merchant wants plentiful and varied winds to keep his ships moving. Not surprisingly, the Water Miller asks for rain while the Wind Miller seeks wind. Likewise, a gentlewoman demands temperate weather in order to protect her fair complexion, but the laundress wants plenty of sun to dry her clothes on the line. Set in pagan Rome, the play nevertheless employs Christian concepts throughout. In answer to Jupiter's question about who he is, the clownish servant Merry Report provides a jesting answer—"I am I" (106), echoing the "I am who I am" of Exodus 3:14. The young boy calls Jupiter "godfather god" (1013) while Jupiter refers to himself as "Principal / god and governor of heaven" and seeks to instill "universal peace" in his response to the many suitors (1246–8). Moreover, in his capacity as a representative of Jupiter, Merry Report suggests Christ as a figure of divine incarnation.

Efforts to stage the divine order of nature found new expression in the reformed cycle plays of the Protestant polemicist John Bale. A Carmelite friar who left the clergy in order to marry, Bale fell under the protection of Thomas Cromwell during the early years of the English Reformation, writing many powerful texts against the Catholic church. Bale shared with his friend and fellow exile John Foxe the hope that, as Foxe declared in his

Acts and Monuments (1563), "players, printers, and preachers" might form a "triple bulwark against the triple crown of the pope."[32] To that end, Bale wrote at least three plays that formed a Protestant "cycle" of biblical drama telling the stories of biblical prophets of the Old Testament (*God's Promises*), the coming of Christ (*John Baptist's Preaching in the Wilderness*) and Christ's victory over Satan (*The Temptation of Our Lord*), all published in 1538. These plays foreground a mode of theatrical representation that is conscious of its own artificial construction, perhaps in order to avoid the charge of idolatry that Bale himself applied to Catholic claims about the eucharist as the real presence of Christ.[33] In the same year, Bale wrote a humanist interlude on the *Three Laws of Nature* (c. 1538) that mixes the rhetorical and ritual structures of the spoken word of preaching with the mediums of drama and print in the service of reforming the church. The interlude begins with a dialogue between "Lex Natura" ("Law of Nature") and the scatological upstart "Infidelitas" (Infidelity) that frames the drama in terms of a divinized nature. Infidelity hopes to corrupt Nature by employing "pestilent idolatry" and "most stinking sodomy."[34] Left alone on stage, he conjures the figures of Sodomy and Idolatria, who vow to help him in his undertaking. Idolatria, who has recently assumed the form of a woman, contrives many "subtleties" that go well beyond traditional forms of religious observation. She uses her holy water, for example, to ruin both corn and cattle, trees and herbs (sigs. C2-C2v). Her goal is to corrupt the heart of all Christians, while Sodomy aims to corrupt the flesh by inspiring them to lusts of all kinds. Suffering leprosy and other diseases of the flesh, as well as perversions of the heart caused by the subtle machinations of Sodomy and Idolatria, Lex Natura calls for mercy and pity at the end of act 2, setting up the arrival of Moseh Lex (Law of Moses), who explains judicial law to Infidelity, who in turn implores Infidelity to refrain from drinking and other forms of "outward injury" (sig. C3). But new vice characters, including Ambition and Covetousness, blind Moseh Lex, establishing the necessity for the arrival of Evangilium, or "Christ's Gospel," who proves stronger than the Old Testament commandments of Moses in his ability to restore faith and forgiveness to the flock (sig. G2). Against the established church of Rome, which he dismisses as a group of "disguised hypocrites," Evangilium posits a "secret Church" based on faith rather than "outward ceremony" (sigs. G3-G3v). Unable to cleanse England of hypocrisy and pseudodoctrinal beliefs, Evangilium gives way to a more vengeful god, Vindicta Dei, who threatens Infidelitas with the kind of punishments he has already performed on godless populations in biblical history, including the inhabitants of Sodom and Gomorrah.

The continued rebellion of Infidelitas invites the final divine figure into the play, Deus Pater, who restores the laws of Nature, Moses, and Christ to health and wholeness. Renewed and restored, the three laws vow obedience to Deus Pater, who urges Christi Lex to "Work thou in the heart, a knowledge necessary / In the flesh" in order to achieve inward transformation by means of "outward Ceremony" (sig. K3v). This movement between outward form and inward spirit shapes many of the plays we have examined. The reformed church must rid itself of idolatry and hypocrisy, the play suggests, but it cannot rid itself of embodied action and forms of "ceremony." Even as Deus Pater dismisses as "outward ceremony" that risks idolatry, he also reminds Christi Lex of the dependence of "inward spirit" on mimesis (sig. G3). At stake once again is the relation between drama as ritual and as mimesis, between representation of the spirit and of the flesh.

These dramatic interludes by Rastell, Medwall, Heywood, and Bale helped to alter the definition of nature by integrating scientific and religious debates with new forms of theater. They established drama as a space for reframing nature in relation to God and to the possibilities of worldly representation. Scholars have rightly rejected narratives of easy or linear evolution from medieval precedents, and it is fair to say that the new modes and genres of drama that flourished in early modern London maintain both continuities and differences from previous traditions. Important differences include staging techniques. Where the altar of the *Quem quaeritis* and other liturgical tropes centered and fixed the gaze of the parishioner, for example, the staging practices in cycle plays and morality plays established a dialectic between a fixed center or *locus* upstage near the tiring house in which characters of high social status such as kings, religious figures, and nobles remained largely stationary and the open space or *platea* where vice characters, fools, and others moved more freely, challenging the fixed authority of their social betters.

According to Robert Weimann, Shakespeare adapted and transformed this structure in his works.[35] Weimann argues that the neutral and undifferentiated *platea* and the fixed *locus* of symbolic action intermingle as spaces in Shakespeare, producing new types of immersive experience for early modern playgoers. Shakespeare's theatrical accomplishment, according to Weimann, derives not only from its reliance on popular tradition, but also from his flexible and innovative approach to formal aspects of drama derived from classical humanism. Shakespeare mixes classical and medieval sources with English traditions of popular festivity to create hybrid forms of theater that combined *locus* and *platea* in new ways (258). As Weimann notes, "Shakespeare's was a dramaturgy of kings and clowns, of 'mongrell Tragicomedie,' a theater in which the localized throne was nonetheless functionally connected with the more downstage *platea* position" (244).

For Weimann, moreover, "the Elizabethan theater did not strive to create a visual illusion of actuality," but it "did attempt to imitate nature, albeit in poetically heightened terms" (216). Weimann also observes that "for the Elizabethan playgoer the drama was more than a play taking place on a stage separated from the audience; it was an event in progress in which good listening and watching were rewarded" and the spectator felt part of the performance (213). Weimann explores, for example, the "double vision" or "principle of multi-consciousness" in the servant plot that connects Elizabethan plays with Roman theater and folk traditions (190), but he has little to say about the traditions of theology and natural philosophy that also shaped medieval and early modern drama. Still, the model of spectatorship that emerges in Weimann's study of Shakespeare and the popular tradition reduces boundaries between performer, spectator, and performance of the *imitatio* tradition. The platform stage and the architectural structure of the early modern playhouse in London mingled various social spaces together, subjecting them to a leveling effect. The new drama performed in permanent, commercial playhouses follows earlier theatrical traditions that constructed forms of mimesis around debates about how to know and interpret the order of nature.

Spectacular Nature in *Doctor Faustus*

One example of such theatrical mimesis produced during Shakespeare's own lifetime is Christopher Marlowe's wildly popular *Tragedy of Doctor Faustus* (c. 1588–92). Performed and revised by at least two London playing companies over several decades, it was one of the most frequently staged and visually spectacular works of drama in the early decades of commercial theater in London. In the prose *History of the Damnable Life and Deserved Death of Doctor Faustus* (c. 1592) that was printed around the same time that Marlowe wrote his play and that he may have used as a source (perhaps by drawing on lost print or manuscript copies), Faustus falls into "fantasies and deep cogitations" after he has achieved the status of Doctor of Divinity, becoming a "Speculator" who rejects scriptures and seeks the "Devilish

Arts" that he finds in Persian, Chaldean, Hebrew, "Arabian," and Greek books of "Necromancy, Charms, Soothsaying, Witchcraft," and "Enchantment."[36] The Faustus in this prose version of the story rejects divinity in favor of the more "worldly" pursuits of astrology, math, and medicine.[37] Marlowe's Faustus, by contrast, expresses a greater desire for control over territory, aspiring to become a "demigod" who will "reign sole king of all the provinces" currently occupied by the Spanish Prince of Parma.[38] But he is also a restless intellectual who rejects traditional disciplines of university knowledge, including medicine, law, and divinity, casting aside books of Aristotle, Galen, Justinian, and Jerome in favor of the "metaphysics of magicians / And necromantic books" that he finds more "heavenly" (B 1.1.48–9) and that offer a world of "profit and delight" (B 1.1.52).

Marlowe dramatizes the rise and fall of Faustus in and through modes of theatrical visuality that respond to questions of natural philosophy and theology. The spectacular staging of the play, which likely included a Hell mouth opening over a trap door and a flying devil propelled by fireworks, made it one of the most visually spectacular works of drama on the early modern stage. Even though the magic that Faustus desires itself depends on its own "necromantic books" that are also "heavenly" (B 1.1.49)—one of which Lucifer himself gives to Faustus in the A text (A 2.3.166)—Marlowe represents the necromancy of Faustus in a variety of verbal, textual, and visual forms. Natural magic must be visualized in the play in order to be understood. Early in the play, for example, Faustus demands of his mentors in magic Valdes and Cornelius that they "show" him some "demonstrations magical" (A 1.1.150). In response, Valdes vows to give Faustus copies of books by the medieval magicians Roger Bacon and Pietro d'Abano while also recommending both the New Testament and the Hebrew Psalter (A 1.1.154–5). Roger Bacon had a reputation as a conjurer in the sixteenth century, but through his magic he also conducted new forms of experimentation on the order of nature.[39] Cornelius emphasizes that Faustus must learn the "words" and "ceremonies" of the "art" of magic before he can try such "cunning himself" (A 1.1.158–60). Marlowe describes Faustus's desire for magic in terms of "lines, circles, signs, letters, and characters" (A 1.1.51 / B 1.1.50) that he must both discern in the order of nature and learn to use himself as a conjurer and interlocutor with devils. His successful conjuration reveals the "virtue" of his "heavenly words," even though such words ultimately produce a devil (A / B 1.3.27).

The magic of Faustus in Marlowe's version of the story is thus a spectacular hodgepodge of natural philosophy, biblical texts, verbal sorcery, and modes of supernatural intelligence. Cornelius describes the "miracles" of magic to Faustus as "grounded in astrology" and "well seen in minerals," for example (A 1.1.136–9). As much as Faustus fantasizes about magic as a means of gaining wealth, power, and influence, his initial discourse with Mephistopheles concerns cosmology and other forms of knowledge that he has claimed are inaccessible to him in the academic fields he has studied. The first question Faustus asks Mephistopheles after signing away his soul to Lucifer concerns the location of Hell in the universe, a question that the devil initially answers flippantly ("under the heavens" [A 2.1.114 / B 2.1.112]) before defining it as a placeless condition experienced by the prideful and fallen. Later, Faustus disputes with Mephistopheles about "divine astrology," asking him how many "heavens" exist "above the moon" and other questions of classical, medieval, and early astronomy (A 2.3.34–5 / B 2.3.32–3). Out of this disputation emerges a question that Mephistopheles refuses to answer—namely who created the earth (A 2.3.64–5 / B 2.3.66). In other words, the line of questioning in which Faustus engages leads him along a path of natural philosophy that ends in a question of divine creation. At stake in the play is not just the spiritual status of Faustus within Christian theology but also the question of who created the *ordo creationis* and whether it was created for mankind. This helps explain why the trope of the Book of Nature becomes visually central in the play when Wagner speaks of Faustus' efforts to learn the "secrets of astronomy" that are "graven in the book of Jove's high firmament" in the Chorus of act 3 (A/B 3.1.1–2). Natural knowledge for both Faustus and the audience is both godly and potentially demonic in Marlowe's treatment of the Faustus myth.

In lieu of answering the question by Faustus about who created the world, Mephistopheles produces a theatrical entertainment featuring grotsequely embodied performances by the Seven Deadly Sins. Mephistopheles urges Faustus to respond as a scholar—to "question them of their names and dispositions" (A 2.3.105–6 / B 2.3.107–8). In this and other modes of extravagant theater in the play, including the richly appareled devils who dance and delight Faustus when his blood congeals as he attempts to sign away his soul (A 2.1.64–70 / B 2.1.62–8) and the devil dressed as a woman with fireworks whom Mephistopheles conjures to dissuade Faustus from seeking a wife (A 2.1.143 / B 2.1.137), the devils provide spectacular distractions rather than scientific knowledge. Marlowe's theater also generates visual spectacles of pleasure that distract Faustus from theological matters. For example, Mephistopheles calls on Lucifer and Beelzebub to turn Faustus away from the Book of Nature and the contemplation of

Christ that it would entail. Lucifer urges Faustus to "Talk not of Paradise or creation, but mark the show" of the seven deadly sins (A 2.3.102–3 / B 2.3.15–6). Likewise, following the appearance of the Old Man who urges Faustus to seek the mercy and grace of God, Mephistopheles bemoans that he "cannot touch his soul" but will "afflict his body" with the temptation of Helen (A 5.1.78–9 / B 5.1.81–2). Her spirit is fully embodied—Faustus can kiss her, and vows to "dwell" in the "heaven" that is her lips (A 5.1.95 / B 5.1.98). For spectators, our Faustian bargain involves both placing ourselves in the position of Faustus as a transgressive seeker of power and becoming complicit through acts of theatrical spectacle that conjoin us with Marlowe's transgressive forms of knowledge.

Like the Towneley *Creation*, moreover, Marlowe does not stage the heavenly cosmos directly. Rather, he allows Faustus to describe it using a language of "tropics, zones, and quarters of the sky" that he glimpses from his vantage point on a chariot drawn by dragons that apparently flies above the earth (A/B 3.0.1–2, 7). This crucial moment, at the exact center of the play, invokes the divine order of nature as an emblem of the highest knowledge Faustus achieves during his time with Mephistopheles. From this point forward in both extant versions of the play, magic loses its relation to natural philosophy and becomes, instead, a means of achieving visual tricks and slapstick humor, such as the jests that Faustus and Mephistopheles play on Pope Adrian and his followers in Rome and the apparitions of Alexander and his paramour that he conjures for the Holy Emperor.

The B text of Marlowe's *Faustus* also contains another example of natural knowledge when Faustus and Mephistopheles travel above Naples, Paris, Venice, and Padua on their way to Rome at the beginning of act 3. Looking out across the world from the vantage point of the heavens, Mephistopheles promises that the holy city will "delight" the "eyes" of Faustus (B 3.1.31). Faustus thanks his guide for providing views of the face of heaven, earth, and hell, noting that his visions of the kingdoms of the world have pleased his "eye" (B 3.1.74), but he begs to be an "actor" in what he has seen rather than just a passive observer (B 3.1.75). Where the theatrical pageant of the Seven Deadly Sins had earlier distracted Faustus from questions of cosmological creation, here the visual spectacle of the planet itself spurs Faustus to desire to join that spectacle as a player.

This visual and embodied fantasy produced by the illusion-making structure of Marlowe's theater ultimately trumps the possibilities of magic for producing new knowledge about nature and even the ambition that Faustus has for magic as a form of power over other nations and peoples. Nevertheless, *Doctor Faustus* frames its central visual spectacles in terms of the desire to know both God and the *ordo creationis*. Marlowe uses words, action, movement, and stage magic to question the epistemological status of the order of nature. One possible witness to this new theater of nature was William Shakespeare, who joined the Lord Chamberlain's Men in 1592 and regarded Marlowe as a friend, rival, and mentor in his own ambitious efforts to become a playwright.[40]

Estranging Nature in *Romeo and Juliet*

As we saw in the previous chapter, Shakespeare's *Romeo and Juliet* joins Marlowe's *Faustus* in framing nature as both a divine and potentially demonic force that can be known in new ways through theatrical representation. Friar Laurence surveys the order of nature, drawing on debates in theology and natural philosophy that shaped the work of Pliny and numerous contemporary herbals that reframed that order in early modern Europe. His effort to read the order of nature is part of a larger structure in the play that situates theatrical representations in and through modes of natural philosophy in order to question the legibility of nature. His praise of the "powerful grace" that stretches across "plants, herbs, [and] stones" also anticipates the dispersal of agency of Romeo and Juliet across the order of nature in the play.

For example, Shakespeare depicts the tragedy and especially the lovers themselves in complex cosmological language. The arc of the sun defines Romeo's character beginning with the first reference to him by Benvolio, who spies him entering a forest just before sunrise (1.1.120–31). In the final lines of the play, moreover, the Prince describes a "glooming peace" descending over the shattered city state of Verona in terms of the sun that hides its head from "sorrow" (5.3.305–6). Celestial metaphors likewise famously define the "star-crossed lovers" themselves (1.0.6), helping Romeo articulate his love for Juliet ("what light through yonder window breaks? / It is the east, and Juliet is the sun. / Arise, fair sun, and kill the envious moon" [2.1.44–6]) and, in its contrast with the night, to describe the paradoxical passion of the young lovers. Juliet describes Romeo as "day in night" (3.2.17) as she wishes for shorter days in order to prolong the nighttime she can spend with him:

> Come, gentle night; come, loving, black-browed night,
> Give me my Romeo; and, when I shall die,
> Take him and cut him out in little stars,
> And he will make the face of heaven so fine

> That all the world will be in love with night
> And pay no worship to the garish sun. (3.2.20–25)

Romeo's allure is so great, she suggests, that, even if he were turned into stars, he would make "all the world" love night rather than the "garish sun." Lord Capulet echoes this alignment when he describes the young women who will attend his party as "Earth-treading stars that make dark heaven light" (1.2.25). Furthering the association, Romeo describes Juliet on first meeting her as "beauty too rich for use, for Earth too dear" (1.4.158). Later that night, just before he climbs the wall into Juliet's garden, he remarks, "Can I go forward when my heart is here? / Turn back, dull earth, and find thy center out" (2.1.1–2). The invisibility of this center contrasts with the more visible firmament in which the lovers situate their love, but it also connects Romeo to the chthonic forms of the order of nature that Friar Laurence invokes. Romeo may unify "heaven" and "earth," as Friar Laurence remarks, but he is also dispersed among these in such a way as to leave him without a center (3.3.120).

Searching for Romeo following the Capulet party, the supernaturally attuned Mercutio offers to conjure a "spirit" in the shape of his mistress, whom he still thinks to be Rosaline (2.1.24), but Benvolio guesses that Romeo has "hid himself among" the trees in the dark night, where he is the consort of "the humorous night" (2.1.30–31). Likewise, Juliet's father describes her as continuous with the "sunset" of his "brother's son," the slain Tybalt, and sees in her "conduit" of rain the "counterfeit" or representation of "a bark, a sea, a wind" (3.5.126–131). Juliet's materiality, in other words, emerges in and through the weather, just as Romeo is a product of the sun, stars, sky, and earth. Although each lover is defined differently in the play, their shared semiotic flexibility binds them together in both love and death. Indeed, their movement between these poles replicates the duality of the divine "grace" of the order of nature that Friar Laurence surveys.

On learning of Romeo's murder of Tybalt, Juliet engages in metaphysical speculation about her lover's creation, invoking demonic forces in the process:

> Despisèd substance of divinest show!
> Just opposite to what thou justly seem'st,
> A damnèd saint, an honorable villain!
> O nature, what hadst thou to do in hell,
> When thou didst bower the spirit of a fiend

In moral paradise of such sweet flesh?
Was ever book containing such vile matter
So fairly bound? O that deceit should dwell
In such a gorgeous palace. (3.2.77–84)

The passage represents Romeo as a divine product of nature—a Book of Nature whose beautiful outside contains a fiendish devil. In this creation myth of Shakespeare's own invention, Nature descends to Hell to create such a creature with a "gorgeous" exterior that conceals more "vile matter." When the Nurse speaks disparagingly of Romeo, however, Juliet turns the tables, cursing her and arguing that Romeo is "Sole monarch of the universal earth" (3.2.94). In both cases, the character of Romeo emerges through shifting descriptions of a natural order, which itself has been reconfigured as a semiotic system.

This central motif of a corrupt interior hidden by a graceful exterior also shapes other observations about love in the play. When Romeo complains in the opening scene that he is out of favor with Rosaline, for instance, Benvolio describes his condition in terms of the irony that "love, so gentle in his view, / Should be so tyrannous and rough in proof" (1.1.164–5). The quick-witted Romeo turns that relationship on its head by arguing that "love, whose view is muffled still, / Should, without eyes, see pathways to his will" (1.1.166–7). Benvolio's description of Rosaline as a stone-cold Petrarchan object of desire who is "tyrannous" for refusing love becomes, in Romeo's revision, a blind Cupid who can make love happen regardless of the "will" of the person he shoots with his arrows of passion. (The dramatic irony of this proposition becomes clear in the following scene as Romeo is himself struck blind with love for Juliet, making him forget all about Rosaline.)

Where Friar Laurence identifies a moral interlude of sin and grace in the order of nature, Romeo sees divinity in Juliet herself, whose body he describes as the product of divine "heaven" (2.1.57, 62, 70). The lovers thus proudly make themselves into objects of their own "idolatry" (2.1.154–6) and inhabit "heaven" as a space of personal intimacy on earth rather than as a site of otherworldly bliss ("heaven is here / where Juliet lives" [3.3.29–30]). In the closing lines of the play, Friar Laurence blames "a greater power" than "we can contradict" for the deaths of the lovers (5.3.153). The Prince supports this account, proclaiming the deaths of Romeo and Juliet a "scourge" of "heaven" (5.3.292–3). But such a providential narrative does not come to fruition in the play itself, as we have seen. If there is a special providence in the play, it seems governed by the hidden virtues of things, not produced by Christian or pagan gods.

The anxieties of representation at the heart of the play focus instead on the ways in which love as a product of the order of nature can be articulated and known. In keeping with the mimetic imperative of forms of early modern philosophy we have been tracing, the world of the play is a world in which men (and women) must "show" themselves. This anxiety of revelation focuses mostly on sexual desire and the manifestation of that desire through embodied action. Benvolio thus offers to "show" Romeo young women at the Capulet party who will make him forget his beloved Rosaline (to which Romeo replies that "no such sight" can be "shown") (1.2.89, 103). The question of how to stage invisible divinity that occupied English and European drama since the *quem quaeritis* is thus reworked in Shakespeare in terms of a discourse of romantic love, which Romeo calls the "devout religion of mine eye" (1.2.91). The play repeatedly invokes such language of vision to express its own complex desire to represent invisible passions and to sway the audience with voice, language, and gesture. The following chapter explores this religion of love more fully as Shakespeare takes up both Petrarchan and Reformation ideas of idolatry in the process of experimenting with new ways to stage language, objects, and bodies in his early comedy of *Two Gentlemen of Verona*. Idolatry emerges in this play as an overdetermined problem of theology, natural philosophy, and representation.

Notes

1. Harrison makes this claim in a range of works, including *The Bible, Protestantism, and the Rise of Natural Science*; *The Fall of Man and the Foundations of Science*; and "The 'Book of Nature' and Early Modern Science."
2. Objections to and revisions of Harrison's claims about Protestant hermeneutics and the rejection of allegorical interpretation include Jitse van der Meer and Richard J. Oosterhoof, "God, Scripture, and the Rise of Modern Science"; Scot Mandelbrote, "Early Modern Biblical Interpretation and the Emergence of Science"; and Richard Serjeanstson, "Francis Bacon and the 'Interpretation of Nature' in the Late Renaissance."
3. E.K. Chambers adopts this evolutionary model, although he also gives ample attention to folk traditions and socioeconomic factors. See *The Medieval Stage*, 2:1–67. Karl Young and Hardin Craig also follow Chambers in their evolutionary claims. For a valuable critique of all three, together with an overview of alternate approaches, see Leonard Goldstein, *The Origin of Medieval Drama*, 13–45.

4. O.B. Hardison, for example, notes that the secular model "assumes that medieval man [*sic*] thought like nineteenth-century man [*sic*], or ought to have done so" in *Christian Rite and Christian Drama in the Middle Ages*, 33.
5. See Thomas Aquinas, *Summa Theologica*, 2415–22.
6. Michal Kobialka, *This Is My Body.*
7. David C. Lindberg, "Science and the Medieval Church," 276. In a different essay, Lindberg notes that medieval philosophers including Aquinas and Albertus Magnus were "so successful" in synthesizing classical natural philosophy with Christianity that "by the sixteenth century Aristotle had taken on the appearance of a Christian saint." See "Medieval Science and Its Religious Context," 77.
8. For an overview of this tripartite scheme of Aquinas and its relation to a range of medieval works of art, including ivories, see Sarah M. Guérin, "Meaningful Spectacles," 66–7.
9. Michael Camille, *Gothic Idol*, 205.
10. For an account of historical debates about divine essence and extension with reference to the extra-biblical doctrine of the Holy Trinity, see Richard Swinburne, "The Social Theory of the Trinity."
11. For a reading of the elevation of the Host as the "most theatrical moment" of church service, together with a reassessment of the claim by E.K. Chambers that the mass was "essentially dramatic commemoration," see T.P. Dolan, "The Mass as Performance Text," 21.
12. Kobialka, 198.
13. See, e.g., the eleventh-century breviary from St. Gall that describes the raising of the host from the sepulchre (*elevation*) reprinted in Young, *Drama of the Medieval Church*, 1:130–1.
14. See, e.g., the eleventh-century text of the trope from Limoges included in *Medieval Drama,* ed. Bevington, 56.
15. For a text of the *Celestial Hierarchy*, see Pseudo-Dionysus, *The Complete Works*, ed. Rorem, 147. See also Kobialka, 209–16.
16. O.B. Hardison and E.K. Chambers both see the ritual of communion as fundamentally dramatic, though from different perspectives. See Hardison, *Christian Rite and Christian Drama* and Chambers, *The Medieval Stage.*
17. I am using the updated text of the Towneley *Creation and the Fall of the Angels* edited for the TEAMS Middle English Text series by Garrett J.P. Epp and produced digitally through the University of Rochester library at https://d.lib.rochester.edu/teams/text/epp-the-creation. I cite this edition, with modernized spelling, hereafter cited parenthetically in the text by line number.
18. As R.W. Hanning notes, drama as a form "peculiarly belongs to God's creatures, not their creator." See "'You Have Begun a Parlous Pleye,'" 29.
19. Auerbach discusses the play, which he calls *Adam and Eve,* in chapter 7 of *Mimesis.*

20. I rely on the edition of the "Service Representing Adam" edited and translated by Bevington in *Medieval Drama*, hereafter cited parenthetically in the text by line number.
21. The biblical story contrasts Abel as a "keeper of sheep" and Cain as a "tiller of ground" (Genesis 4:2), suggesting a broader contrast between shepherds and farmers, but the *Ordo Repraesentationis* does not follow this logic.
22. See, e.g., Bevington's chapter on "Medwall and the New Tudor Ruling Class" in *Tudor Drama and Politics*, 42–53, and R.C. Jones, "The Stage World and the Real World in Medwall's *Fulgens and Lucrece*."
23. Debate continues over the degree and nature of Medwall's innovation in the play. Bevington suggests that the play "borrowed from the popular tradition rather than being part of it" in *From "Mankind" to Marlowe*, 43–44. Charles Read Baskerville likewise regards the work as derivative in "Conventional Features of Medwall's *Fulgens and Lucrece*." Joel Altman regards the play as an innovative exercise in staging classical debate structures as part of early modern humanism in *The Tudor Play of Mind*, 13–32. See also Aaron Kitch, "Medwall's 'Condycion,'" 8.
24. John Rastell, *Nature* (1534), reprinted in *Recently Recovered "Lost" Tudor Plays*. Hereafter cited parenthetically in the text by page number.
25. Liza Blake, "Allegorical Causation and Aristotelian Physics in Henry Medwall's *Nature*," 343.
26. For bibliographical details about both plays, as well as others discussed in this section, see W.W. Greg, *A Bibliography of the English Printed Drama to the Restoration*, 1:81–97; L.B. Wright, "Notes on *Fulgens and Lucrece*"; and J.S. Colley, "*Fulgens and Lucrece*: Politics and Aesthetics."
27. Richard Southern, *The Staging of Plays before Shakespeare*, 57.
28. My interpretation of Rastell's printer's device draws on E.J. Devereux's remarks in *A Bibliography of John Rastell*, 51.
29. John Rastell, *Four Elements*, in *Three Rastell Plays*, ed. Axton, 31, lns. 1, 9, 25. All quotations from this play are hereafter cited parenthetically in the text by line number.
30. I use the text of *Wit and Science* included in the anthology of *Recently Recovered "Lost" Tudor Plays*, ed. Farmer, which does not contain line numbers. I therefore cite this play parenthetically in the text by page number.
31. John Heywood, *The Play of the Weather*, in *Tudor Interludes*, ed. Happé, 143 (lns. 22–8).
32. John Foxe, *Acts and Monuments*, 1348.
33. On Bale's Protestant hermeneutics, see Katharine A. Gillen, "From Sacraments to Signs." See also David Scott Kastan, "'Holy Words' and 'Slypper Wit.'"

34. John Bale, *A New Comedy or Interlude Concerning the Three Laws of Nature*, sig. C1. Hereafter cited parenthetically in the text by signature number.
35. Robert Weimann, *Shakespeare and the Popular Tradition*. Hereafter cited parenthetically in the text by page number.
36. Anon, *The History of the Damnable Life and Deserved Death of Doctor John Faustus*, ed. Rose, 66–7.
37. Ibid., 67.
38. Christopher Marlowe, *Doctor Faustus*, B-text (1.1.60) and 1.1.91–2, ed. Kastan, 58–9. All quotations from this play are to this edition, hereafter cited parenthetically by act, scene, and line number, with "A" designating the A-text and "B" designating the B-text of the play.
39. The author of treatises on astrology, alchemy, and medicine, Bacon likewise engaged in studies of grammar and logic while disseminating important works of Greek and Arabic natural philosophy. See Stuart C. Easton, *Roger Bacon and His Search for a Universal Science* and *Roger Bacon's Philosophy of Nature*, ed. Lindberg.
40. For example, Shakespeare alludes directly to Marlowe's death in a "little room" in *As You Like It* (3.3.9–12) and revises many of Marlowe's plots and characters in his plays. See Patrick Cheney, *Shakespeare, National Poet-Playwright*, 43–4. For a more recent examination of the vexed relationship between the two poet-playwrights, see Robert Sawyer, *Marlowe and Shakespeare: The Critical Rivalry*.

CHAPTER 4

The Nature of Idolatry in *Two Gentlemen of Verona*

Early modern discoveries of nature as a complex structure of representation in both printed books and performed drama enacted a shift from medieval forms of *imitatio* organized vertically between an unseen God and the visible world to new and more horizontal networks of associations and knowledge practices, with striking implications for knowing and representing the *ordo creationis.* Francis Bacon's revolutionary efforts to survey, reform, and restructure the state of "learning" in England illustrate this shift. Bacon grounded his project of inductive philosophy on what he called the "light of nature and experience, which is eternal."[1] At the same time, he focused on the forms of mediation between the scientific investigator and the object of investigation. He was particularly determined to reform the art of rhetoric in order to make it better serve scientific reason. Unlike Aristotle, who defined rhetoric mostly in terms of the operation of judicial proofs, Bacon valued rhetoric as a tool of rational inquiry into the *ordo creationis.*[2] Despite his oft-cited dismissal of overly exuberant humanists who "hunt more after words than matter" in the first book of *The Advancement of Learning*, Bacon sought to reform rhetoric and grammar as tools of scientific inquiry, urging more disciplined attention to the grammar of English as a native tongue and to the "Philosophical" examination of the "power and Nature of words" as the "footsteps and prints of Reason."[3]

In the process, Bacon reimagined language as both a mode of mimesis and a tool of scientific knowledge, especially in his attention to theater as

A. Kitch, *Shakespeare's Theater of Nature*, Palgrave Studies in Literature, Science and Medicine,
https://doi.org/10.1007/978-3-031-78082-0_4

a mode of representation. By defining rhetoric as the means of applying "Reason" to "Imagination" in order to "move the will," for example, Bacon linked rhetoric or "the Art of Eloquence" to the Imagination as a mental structure that adjudicates between appearances and reality.[4] He also praised the category of "representative" poesy as a "visible History" and "image of actions as if they were present."[5] Like Sidney, Bacon maintained a Platonic skepticism about the truth value of mimesis. And yet, also like Sidney, he endorsed forms of "poesy" that produce more "goodness" and variety in the soul of mankind than can be found in the order of nature.[6] The 1640 edition of *The Advancement of Learning* includes a passage on "dramatical or representative" poesy that praises classical drama while denigrating the mere "pastimes" of contemporary culture. Bacon singles out forms of collective witnessing that take place in a theater as "a secret of nature that the minds of men are more patent to affections, and impressions, Congregate, than solitary."[7] For Bacon, therefore, "well-directed" and enlightened theatrical performances have the capacity to neutralize the false modes of enchantment that he calls "idols." Like certain forms of inductive philosophy or embodied rituals of faith such as fasting, the regulated and sense-based knowledge of witnessing the performance of live theater can remedy the inherent intellectual fallacies of human understanding.

Bacon likewise defined language in visual terms in *The Advancement of Learning*. For him, words are "images of matter" and so to fall in love with them is to "fall in love with a picture."[8] After Bacon's death, his follower John Webster extended this argument to claim that "we all study, and read too much upon the dead paper idols of creaturely-invented letters, but do not, nor cannot read the legible characters that are only written and impressed by the finger of the Almighty." For Webster, as for Bacon, we must learn to read the "legible and indelible letters speaking and sounding forth [God's] glory," which he describes "not as mute statues, but as living and speaking pictures, not as dead letters, but as preaching *Symbols*."[9] Despite his skepticism toward humanist *copia* and some examples of imaginative literature, Bacon describes errors of knowledge and judgment in terms of idolatry as a way to acknowledge the necessity of representation and the foundational structure of theology in his new science.

The category of idolatry thus illustrates the interdependent modes of science and religion in Bacon's scientific project. Despite his warning against the "unhealthy mingling of divine and human" in *The New*

Organum, Bacon's ambitious goal reforming knowledge was deeply responsive to religious ideas and practices.[10] In *The New Organum*, for example, he urged natural philosophers to seek "the true vision of the footsteps of the Creator imprinted on his Creatures."[11] Adopting the logic of the Book of Nature, he observed that the "Ideas of the divine" could be found in "the true prints and signatures made on the [works of] creation."[12] And in *The Advancement of Learning*, he claimed that natural "creatures" reveal the power of Nature as a divine "carver."[13] Following Luther and Reformation debates about idolatry, however, he dismissed religion that appealed to the imagination rather than to reason by means of "similitudes, types, parables, visions, [and] dreams."[14] At the same time, he acknowledged the imagination as a structure of representation necessary for all modes of knowledge, including the modes of scientific "invention" to which he was devoted.

Like Galileo, Bacon embraced technological innovations such as the telescope and printing press as hybrids of nature and art that were necessary aids for advancing knowledge about the natural world. In the *New Organum*, for example, he observes that all mechanical arts are "founded on nature and the life of experience."[15] He expanded his description of creation in *Description of the Intellectual Globe* (*Descriptio Globi Intellectualis*, 1612) to include a passage on the human capacity to make "the appearance of a rainbow on a wall by the sprinkling of water." This illusion depends on the mimetic capacity of nature itself, "just as much as when the same effect is produced in the air by a dripping cloud; and on the other hand when gold is found pure in sands, nature does the work for herself just as much as if it were refined by the furnace and human appliance."[16] Because art and nature are so intimately connected, Bacon recommends that art itself should be included as one of the three central categories of natural history. Indeed, for Bacon, the "artificial does not differ from the natural in form or essence."[17]

Yet the human mind for Bacon, despite its divine origins, faced significant impediments to clear and full understanding of the order of nature. Seeking to enhance the capacity of human judgement, Bacon compares the mind to an "enchanted glass" or mirror that is "full of superstition and imposture."[18] In the *New Organum*, he divides inherent human "illusions" into four categories using a language of idolatry as simultaneously theological, scientific, and ludic. Thus the "Idols of the Tribe" describe the innate human elevation of our perception over the "nature of things," favoring the "uneven mirror" of human understanding over the objective order of nature. The "Idols of the Cave" reveal a specific defect of individual understanding that "fragments and discolors the light of nature." The "Idols of the Marketplace" denote the tendency of mankind to favor words over understanding, leading us into "countless empty disputes and fictions," while the "Idols of the Theater" refer to dogmatic and systematic philosophies created like the "false and fictitious worlds" of stage plays.[19] In each of these forms of deception, Bacon locates a struggle between God-given potential and the distorted knowledge of fallen humankind. Errors accrue because of the inevitable biases and misperceptions of our postlapsarian condition. But Bacon believed that new modes of inductive philosophy aided by technology could overcome natural biases and errors of perception, leading to clearer knowledge of the order of nature. Such a process would allow natural philosophers to understand the order of nature on its own terms as a product of divine creation, without falling prey to the misguided mediation of human language.

Like Bacon, Thomas Browne in his *Vulgar Errors* (*Pseudodoxia epidemica*, 1646) also defined idolatry as an obstacle to comprehension of the natural world because it collapses the "sign" into the "thing signified." Browne includes several examples in which humans substitute their own language and ideas for the "substance and thing itself," ranging from the false god Belus to the Catholic doctrine of transubstantiation.[20] To commit idolatry is to fall under the same error of logic for Browne as mistaking the particular for the universal or the part for the whole (116). In addition, the inherent deficiency of understanding represented by idolatry can be "further impaired" by human appetite, that "irrational and brutal part of the soul, which lording it over the sovereign faculty, interrupts the actions of that noble part, and chokes those tender sparks, which Adam hath left them of reason" (8). Yet the status of images remains complex and uncertain in Browne. Even as he derides the emblematic understanding of animals as found, for example, in medieval bestiaries, he still relies on the pictorial tradition as evidence of his natural philosophy. Thus, to take one example, he cites as proof of the conventional claim that salamanders can survive inside fire the visual "testimony" of images of such salamanders in Egyptian hieroglyphics (267). Elsewhere, however, he questions the validity of hieroglyphic knowledge, as in the example of the pelican plucking its own breast to feed its children (367). The final book of *Pseudodoxia epidemica*, in fact, debunks the representation of animals derived "emblematically" from images and popular conceits, dismissing the popular notion that dolphins have curved spines as a "fallacy in vision" that makes us believe that they are curved because they move so quickly (369).

Browne seeks a firmer epistemological foundation than the popular superstitions such as those promoted by the "Idolatry of Maimonides" that the appearance of a fox presages bad luck (405). His repeated references to idolatry concern not just the question of how much faith to put in the emblematic tradition, but also to the traditional problem of how to know and represent God that we have been tracing in this book. Browne dismisses the "Anthropomorphites," for instance, who represent God in the shape of an "old man." He distinguishes between the more "tolerable" images of divinity expressed by the Egyptians in the form of "sacred letters" because they do not pretend to be "corporeal representations." The Egyptians, in contrast with the Anthropomorphites, do not "misconceive" of the representations as corporeal realities (410). Christians who engage in such misconceptions are more damaging and have less of an excuse. A better example of the actual visual existence of God for Browne is the spherical sun, which offers a convincing declaration of the "wisdom of the Creator" in the way that it demarcates the four seasons and differentiates between night and day (423). The revolution of the sun and its movement through the heavens, not to mention its extraordinary power to light the entire earth at once, illustrate the majesty of God.

Browne does not regard Egyptian hieroglyphics as idolatrous because he sees them as natural symbols akin to Adamic language in their capacity to be a "language of things": "for they assumed the shapes of animals common unto all eyes [...in the form of a] primitive form of writing" that "might Adam well have spoken, who understanding the nature of things, had the advantage of natural expressions" (403).[21] For Browne, hieroglyphs communicate only visually, not aurally, and therefore succeed in conveying the "nature of things." Applying the logic of such language to his own period, Browne followed Galileo in identifying mathematics as the actual language of nature, but Browne retains the older concept of natural hieroglyphics that regards the bond between signifier and signified as guaranteed by God.[22] Idolatry for Browne thus mistakes artificial language for natural and therefore misconstrues the nature of God's creation. Both Browne and Bacon also focused on the classical etymology of idolatry. The word derives from *eidola,* which was used by Epicurus and Lucretius to understand the failure of the senses.[23] It also overlaps with the Greek *eidolon*, meaning ghost or "spirit-image." Because there was no Hebrew word for idolatry, moreover, the Greek authors of the Pentateuch combined two words—*eidōlon* ("likeness" or "image in the mind") and *latria* ("worship")—to form a new and foreign compound, *idolatria*. The biblical word for idolatry is therefore itself an artificial construction of language that mixes theological, philosophical, and aesthetic categories.

The word could also be used to describe dramatic representation in medieval and early modern Europe. An anonymous manuscript from the fourteenth century, for example, divides the popular word for playacting, *ludo* ("I play"), into four related categories: storytelling (*narrare*), deception (*decipere*), playfulness, and idolatry (*ydola*).[24] This combination of terms makes us rethink our understanding of drama and other forms of representation in medieval and early modern Europe. It also provides a new way of understanding Shakespeare's fascination with idolatry as a subject throughout his poems and plays. Perhaps the most famous example of such idolatry in all of Shakespeare's plays is the statue scene at the conclusion of his late play *The Winter's Tale*, in which an apparently lifeless statue of Hermione returns to life with the help of music.[25] But Shakespeare explores idolatry in many other works, including *The Two Gentlemen of Verona*.[26] In this early comedy—perhaps the earliest that he wrote for the stage—idolatry is both a central idea in terms of how to understand the "nature" of love and a problem of how theater itself employs language, bodies, and images to as forms of mimesis. The play invokes idolatry repeatedly in ways that link the theater to problems of knowing nature that Bacon describes, as well as to a long history of religious ideas around the worship of God.

Idols of Faith

Idolatry for early modern intellectuals such as Bacon and Browne was a problem of knowledge that was central to new efforts to know and understand the divine order of nature. Their appropriation of idolatry for natural philosophy drew on a lengthy tradition of theology that understood idolatry in terms of both visuality and textuality. In Exodus the Hebrew word for the tool that Aaron uses to carve the golden calf is "*cheret*," meaning an instrument for inscribing both words and images. The golden calf constructed by the Israelites illustrates this principle perfectly, since the artificial object violates the *logos*, the Word of God, by redirecting it toward artificial creation. Patristic and scholastic commentators made further distinctions among images and objects in later centuries. Augustine, for example, warned that even to call God "ineffable" is to risk verbal idolatry, while theologians such as Hildegard of Bingen and Joachim of Fiore actively explored the potential of images to embody divinity, especially in relation to the Holy Trinity—Hildegard mainly through human figures, Joachim in relation to geometrical *figurae* such as interlocking circles illustrating the unity of the Father, Son, and Holy Ghost.[27] As we saw in the previous chapter, Aquinas later made a distinction between the kind of reverence due to the cross (*latria*) with that due to representations of the human Jesus (*hyperdulia*) and images of the saints (*dulia*). Other Christian authors invoked idolatry in relation to questions about how divinity becomes manifest in the world.

As Paul observes, even though pagans could witness the inscription of God's power in the heavens, they could not read it correctly. Such pagans, according to Paul, became "vain in their imaginations, and their foolish heart was full of darkness" (Rom. 1:20–21). Paul's anxiety over the interpretation of idolatry reflects a paradox of sorts within Christianity as a fundamentally iconic religion. For John of Damascus, for example, Christ in his capacity as the living presence of God was arguably also the "first icon—according to nature and without any parallel—of the invisible God is the Son of the Father." According to Jean-Luc Marion, such an argument suggests that "Christ determines the rules of every icon and its conditions of validity."[28] The idea of Jesus as the manifestation of God's "word" (John 1:1) revises the problem as it justifies God's presence in signs.

New interest in idolatry in Europe during the Reformation revived these ancient debates about idolatry and its relation to representation. Protestant efforts to reform religious institutions and rituals invoked idolatry in their critiques of established religious practices, with radical implications for understanding language and other systems of representation in the period. The Reformation also put new pressure on Christians to discern truth from falsehood in order to discriminate between the true signs of Christ and their false mirrors in the Antichrist. Martin Luther acknowledged the inevitable crossover between images and written and spoken words: "for whether I want to or not, when I hear the word of Christ, there delineates itself the picture of a man who hangs on the cross."[29] Because it is no "sin" to retain images of the savior, Luther argued that religious images were "neither evil nor good."[30] But his senior colleague at the University of Wittenberg, Andreas Bodenstein von Karlstadt, was more vehemently opposed to any object or image that interfered with religious spirituality. Karlstadt's iconoclasm, which angered Luther enough to come out of hiding to try to denounce it in 1522, spread quickly across Europe, first to Zwingli in Switzerland and then to Calvin in France, whose version of Protestantism was adapted by Elizabeth for the English church.[31]

The English Protestant and translator of the bible William Tyndale criticized Catholic hermeneutics, referring to Catholic icons such as the shepherd's hook and the Bishop's cross as "false signs" prophesized by Christ and Paul.[32] Against such imposter signs of the false church, Tyndale endorsed the true signs of God's faith as registered both in the sacraments and in the marvels of nature. Tyndale's *Obedience of a Christian Man* (1528) thus distinguishes between the true signs of Christ that "speak" and the false signs of the antichrist, which are silent but excessive in their visuality.[33] Other Protestants employed the accusation of idolatry not just to challenge traditional religious practices, but also to change the hearts and minds of Christians who, they argued, had been falsely seduced by Catholic images and objects.[34]

The Protestant war on images in the sixteenth century led some to understand religious images such as Christ crucified as signs of "love and commemoration," as Henry Holland describes them in *The History of Adam* (1606).[35] Railing against the ornamental shrines of the saints and the vibrant visual representations of Mary and other religious figures, other Protestants aimed to redirect the eyes of Christians toward the Word. A homily "Against [the] Peril of Idolatry"—probably written by Bishop John Jewel and featured prominently in the second book of homilies produced by Elizabeth's court in 1563—emphasizes the unknowability of the invisible, spiritual essence of God:

> For how can God, a most pure spirit, whom man never saw, be expressed by a gross, bodily, and visible similitude? How can the infinite majesty and greatness of God, incomprehensible to man's mind [...] be expressed in an infinite and living image? How can a dead and dumb image express the living God?[36]

Such efforts to channel divinity through language rather than the paintings, engravings, and sculptures inevitably raised additional questions about the status of these modes of representation.

The historian Lawrence Stone has argued that "Europe moved decisively from an image culture to a word culture" in the sixteenth century.[37] Patrick Collinson has also described a shift from "iconoclasm to iconophobia" during this period, yet it is important to note that iconoclastic acts could be both creative as well as destructive.[38] Protestants were not against all images, even as they moved away from medieval definitions of religious images as *libri laicorum*, or books for the laity.[39] As Tessa Watt has noted, moreover, many "godly" households in England employed images in their houses as expressions of piety.[40] In this sense, images that were adequately reformed could be tools of redemptive morality, as in the example of early modern emblem books.[41] Yet intense interest in idolatry in the period highlighted new questions about the words, images, and bodies as objects of worship and as modes of representing the divine.

Idolatry was not merely a theoretical concern. It informed actual worship practices and even challenged relations between images, words, and objects for everyday parishioners. One famous example in England surrounds early efforts to introduce Protestantism to England under Henry VIII. In January of 1538, Thomas Cromwell instructed his agent Geoffrey Chamber to visit the Rood of Grace at the Abbey of Boxley near Maidstone in Kent, where Chamber inspected and seized the large crucifix that had inspired "great veneration" in visitors for centuries. (Henry VIII himself visited the site as part of his coronation celebrations.) Chamber recounts in his letter to Cromwell that he discovered "certain engines and old wire, with old rotten sticks in the back of the same, that did cause the eyes of the same to move and steer the head thereof like unto a lively thing; and also the nether lip in likewise to move as though it should speak." To bring the deceitfulness of this animatronic object to light, Chamber and his men seized the Rood and carried it to the marketplace in the nearby town of Maidstone, where they displayed it as an example of "false, crafty, and subtle handling" by the monks.[42]

But this was not the final destiny of the Rood. Historical records suggest that it was transported to London, where it was the centerpiece of a sermon against idolatry given by John Hilsey, Bishop of Rochester, at Paul's Cross on February 24, 1538. A minister from Maidstone, John Hoker, recounts in a letter written to Heinrich Bullinger in May of that year how the Rood met its final destruction:

> Then when the Preacher began to wax warm, and the Word of God to work secretly in the hearts of his hearers, the wooden trunk was hurled neck-over heels among the most crowded of the audience. And now was heard a tremendous clamor of all sorts of people [...] he is snatched, torn, broken in pieces bit by bit, split up into a thousand fragments, and at last thrown into the fire; and there was an end of him.[43]

In describing the "wooden trunk" as a male body with a "neck" and heels, Hoker reconstructs the disgraced icon of the deceitful rood into a performing body. At Maidstone, the Rood was subject to the disenchanting forces of public protest, while in London the sermon at Paul's Cross reanimated the icon through language and personification. The iconoclastic violence against the Rood thus included its reanimation through spoken and written language, as well as gestures.[44]

State-sponsored acts of iconoclasm like the destruction of the Rood of Grace may have done little to win the hearts and minds of English subjects for the Protestant cause, but they did renew questions about representation and modes of worship. Zwingli had warned against the inherent human desire to materialize God, but Calvin elevated the human body as a source of divine power, urging Christians to "cast their eyes upward or downward" in order to note the works of God, including humans created in His image.[45] In a sermon against idolatry printed in England in 1561, Calvin remarks that our "likeness of God" extends to the whole body, not just the mind, heart, and soul.[46] Calvin's *Institutes of Christian Religion* (*Institutio christianae religionis*, 1536) understands idolatry in terms of any effort by Christians to employ bodily senses as tools of apperception, including the Catholic communion, which represented Christ as body without soul and therefore should be rejected as a "detestable idol."[47] Calvin therefore rejected all attempts to create images of God, whose essence he believed to be beyond the capacity of human discernment. At the same time, Calvin encouraged readers to employ their senses of sight and hearing in order to bear witness to the divinity of nature. He describes nature in *The Institutes*, for example, as a "beautiful stage [*pulcherrimo theatro*] to take a godly delight of the manifest and ordinary works of God" (sig. G2v).[48] Where lifeless objects often lead to idolatrous worship for Calvin, the human body is alive to the touch of God and therefore participates in "His glory." Such bodies imbued with God's grace were powerful antidotes for some Protestants to the "dead" images that distracted Catholics from the true objects of divine worship.[49]

For Calvin, the order of nature included trees, rocks, and rivers as well as works of art. He compared the divine order of nature to "a large and gorgeous house furnished and stored with abundance of most finely chosen stuff" (G2v). This analogy pictures God as the chief scenic designer of a cosmological drama in which Christians are both actors and spectators. Yet his resulting model of divine spectatorship is fully immersive, encouraging us to "consider that God hath ordained all things for our guard and safety" and "therewithal do feel His power and grace in ourselves." Such meditative observation will "thereby stir up ourselves to the trust, invocation, praise and love of Him" (G3). Despite the ubiquity of God's presence in the Book of Nature, however, Calvin highlights the difficulty of humans, steeped in original sin, to understand nature as a form of divine revelation.[50]

Protestant critiques of transubstantiation as a cannibalistic act of idolatry also participated in a new hermeneutics of grace with wide-ranging implications for early modern drama, as we saw in the previous chapter.[51] Where transubstantiation identifies the literal presence of Christ in the wine and bread of the sacrament, Protestant consubstantiation interpreted communion as a form of commemoration or memorialization of Christ's promise to his followers and relocated the "presence" of God in the word of the bible itself and in the Book of Nature.[52] For Luther, such relocation of presence did not elevate language in and for itself so much as work to unify or "fasten" (*faast*) spirit to matter and words to things.[53] The charge of idolatry thus had the effect of relocating and dispersing the presence of God more broadly, just as the Lutheran doctrine of "ubiquity"—which attempted to defend the substantive reality of the eucharist—insisted on the omnipresence of God in created matter.[54] Luther's rejection of traditional structures of ecclesiastical hierarchy promoted the view that all vocations could be spiritual.[55] Such claims were simultaneously practical and theoretical, material and symbolic. Sixteenth-century authors in England shifted iconicity toward language, which functioned as a reformed picture in the sense that the image within the word was contained and redirected toward the invisible essence of God.

It is not surprising, given the above claims, that concepts of idolatry helped to define new modes of theater in early modern England. Allusions to idolatry in the context of dramatic mimesis can be found as early as the manuscript known today as the *Tretise of Miraclis Pleyinge* (c. 1400), which criticized plays for their tendency to "delight men bodily" instead of spiritually.[56] Yet the *Tretise* also defended theatrical "personation" or acting that produced a living or "quick" as opposed to dead "book." Elizabethan antitheatricalists developed much more elaborate descriptions of the newly permanent theaters in their published writing. The pamphleteer Philip Stubbes, for example, claimed that the early modern theater promised only to "nourish us in idolatry, heathenry, and sin" by mixing "scurrility with divinity."[57] Likewise, William Rankins condemned playwrights in his *Mirror for Monsters* (1587) whose plays consecrate "Temples of Idolatry" such as those dedicated to Mars and Venus. He dismissed the early modern theater itself as "imperfect, polluted, and defiled."[58] For Stephen Gosson, moreover, early modern theater produced "idolatry" analogous to those pagans who consecrated temples to their gods, especially Venus. According to Gosson, spectators do not innocently absorb such desecration, but are rather, like Faustus, "ravish[ed]" by pleasurable spectacles that render them "carnally minded."[59] For Anthony Munday, moreover, the commercial theaters of early modern London were spaces of diabolical communion in which the "unsavory morsels of unseemly sentences passing out of the mouth of a ruffian player" seems to content the "rude multitude" even more than the legitimate eucharist during communion. Even though the audience is "always eating" such words, they are "never satisfied" by them because those words are empty of substance.[60]

"Engraven Strife with Nature"

Early modern definitions of idolatry as a scientific, religious, and mimetic category also shaped one of the most famous images of William Shakespeare created in early modern England: the engraving by the Anglo-Flemish artist Martin Droeshout printed on the frontispiece of the playwright's collected works in 1623 (see Fig. 4.1). Apparently reluctant to feature a large portrait of Shakespeare without an explanatory text, the compilers of the first folio, John Heminge and Henry Condell, included a commendatory poem by Ben Jonson on the opposite page.

> This Figure, that thou here seest put,
> It was for gentle Shakespeare cut;
> Wherein the Graver had a strife.
> With Nature, to out-do the life:
> O, could he but have drawn his wit.
> As well in brass, as he hath hit.
> His face; the Print would then surpass.
> All, that was ever writ in brass.
> But, since he cannot, Reader, look.
> Not on his Picture, but his Book.

Describing playful competition between the engraver and "Nature," Jonson's poem redirects the reader's gaze from "picture" to "book,"

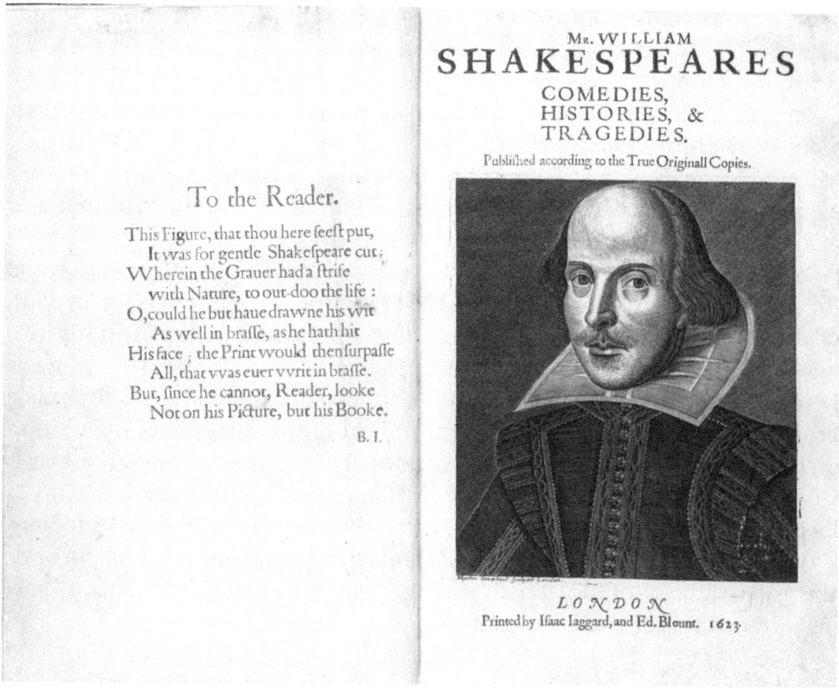

To the Reader.

This Figure, that thou here seest put,
It was for gentle Shakespeare cut;
Wherein the Grauer had a strife
with Nature, to out-doo the life :
O, could he but haue drawne his wit
As well in brasse, as he hath hit
His face ; the Print would then surpasse
All, that vvas euer vvrit in brasse.
But, since he cannot, Reader, looke
Not on his Picture, but his Booke.

B. I.

Mr. WILLIAM
SHAKESPEARES
COMEDIES,
HISTORIES, &
TRAGEDIES.
Published according to the True Originall Copies.

LONDON
Printed by Isaac Iaggard, and Ed. Blount. 1623.

Fig. 4.1 William Shakespeare, *Comedies, Histories, and Tragedies* (1623), title page. (Credit: World Digital Library)

mobilizing a Neoplatonic view of the ultimate "reality" of Shakespeare that disembodied "wit" can only approximate. Such concerns also appear in Jonson's commendatory poem in the folio, which claims that "Nature herself was proud of his designs, / And joy'd to wear the dressing of his lines."[61] Like Hamlet in his instructions to the players, Jonson understands Shakespeare's plays as a performance of nature.

Jonson also invokes Protestant approaches to idolatry in urging viewers to turn away from Shakespeare's "Picture" in order to read his "Book." Like his fellow playwrights, Jonson suggests that language and embodied action combine on stage to represent nature and situates the portrait of Shakespeare of the first folio in terms of the problem of representing that nature. Jonson's playful invocation of the problem of idolatry also hints at its centrality in Shakespeare's works, especially in the context of eroticized adoration. In *Venus and Adonis* (1593), for example, Venus chides Adonis as a "Well-painted idol, image dun and dead" and a "lifeless picture [....] contenting but the eye alone" after he rebuffs her advances (lns. 211–13). Likewise, Lysander justifies his right to pursue Hermia in the opening scene of *A Midsummer Night's Dream* (c. 1595) by noting that Demetrius, whom Hermia's father favors, has been in a relationship with another woman, Helena, who "devoutly dotes…in idolatry" on the "spotted and inconstant man" (1.1.109–10). Helen confesses her own "idolatrous fancy" for Bertram early in *All's Well that Ends Well* (1.1.94), while Berowne in *Love's Labor's Lost* (c. 1594) satirizes Longueville for making "flesh a deity, / A green goose a goddess: pure, pure idolatry" in his love for Maria (4.3.69–70). These examples draw on the widely popular Petrarchan tradition that links falsity of appearances—and especially female beauty—with the falsity of idols, but Shakespeare also filters his writing through Protestant attitudes toward idolatry as a condition of the heart. Consider, for example, the passage in *Romeo and Juliet* when Juliet urges Romeo to swear "by thy gracious self, / Which is the god of my idolatry, / And I'll believe thee" (2.1.155–7). Juliet both critiques and reappropriates the language of idolatry in urging Romeo to turn his idolatrous passion inward rather than outward.

Like Petrarch, Shakespeare's language of idolatry combines erotic and religious registers. But where idolatry for Petrarch is a principle of poetic autonomy, as John Freccero notes in his influential essay on the topic, for Shakespeare it also addresses questions about theatrical mimesis.[62] Idolatry foregrounds problems of physical presence and absence that, as Richard Wilson has claimed, also shape early modern theater as a medium.[63] Idolatry frames Shakespeare's technical approach to crafting stage pictures, for example, in the early comedy of *Two Gentlemen of Verona*, which also relies on familiar associations between idolatry and the central problem of how to know and represent erotic love. The opening scene relies on the language of religious pilgrimage to depict the homosocial friendship between the aspiring courtiers Valentine and Proteus:

> PROTEUS
> Wilt thou be gone? Sweet Valentine, adieu.
> Think on thy Proteus when thou haply seest
> Some rare noteworthy object in thy travel.
> Wish me partaker in thy happiness
> When thou dost meet good hap; and in thy danger,

> If ever danger do environ thee,
> Commend thy grievance to my holy prayers,
> For I will be thy beadsman, Valentine.
> VALENTINE
> And on a love-book pray for my success?
> PROTEUS
> Upon some book I love I'll pray for thee. (1.1.11–20)

Proteus recommends that Valentine find some "rare noteworthy object" in his travels as a token of their absent friendship, while Proteus offers to serve as a "beadsman" to his friend, a reference to the traditional Catholic practice of hiring a person to say prayers on a rosary for another. Valentine also mixes religious and amatory metaphors in his reference to a "love-book" on which Proteus will pray for Valentine's "success," establishing a link that will repeat itself throughout the play.

Valentine's impending absence prompts anxieties of presence in Proteus, for instance in the wished-for "noteworthy object" that will substitute for the absent Proteus and then in the holy bead or book when Valentine encounters "danger." The erotic desire of Proteus for Julia and the anticipated attraction of Valentine for some unknown "object" of his affection help to establish the erotic economy of the comedy. This initial dialogue imagines love not just as a force of personal transformation, as in Ovid's *Metamorphoses*, but also as an unobtainable goal that necessitates a series of substitutions for the unrepresentable thing itself. The objects that might bond Valentine and Proteus together despite their absence can only achieve presence in the play through linguistic description. Indeed, much of the play foregrounds often failed attempts to understand love in terms of words, objects, and images.

Valentine and Proteus serve, in this sense, as Baconian knowledge workers in the play. They foreground experience over precepts and they seek new forms of knowledge grounded in the art of rhetoric. They are ambitious young courtiers who must learn the methods of language to gain love and power. Their erotic conquests and their elevation in social status at the court of Milan depend on their ability to understand and master sophisticated forms of language. Yet they must also learn to match their outward behavior with their inner desires. The violent climax of the play brings these elements into focus as Proteus threatens to seize Sylvia by force:

Nay, if the gentle spirit of moving words
Can no way change you to a milder form,
I'll woo you like a soldier, at arms' end,
And love you 'gainst the nature of love—force ye.
[*He seizes her*]. (5.4.55–8)

This threat of physical violence follows failures of language, revealing the inability of Proteus to master the language and action of courtly love. We will return to this disturbing and climactic scene later in the essay, but for now I address it for the way it shows how questions about how to conquer love mix with problems of language, desire, and representation.

The female protagonists of *Two Gentlemen,* Julia and Silvia, echo their male counterparts in questioning the relative status of words and images as vehicles of desire. Likewise, the courtly fool Speed uses complex wordplay and chopped logic to mock the kind of eloquent wit that Valentine and Proteus aspire to as courtiers-in-training. In his repeated insistence on the blindness of love, together with his skewering of romantic conventions held by Valentine (and mocked together with Lance in their itemizing of the attributes of the milkmaid in 3.1), Speed enacts a mode of satire that tears down pretense but offers nothing of substance in its place. His language is disembodied and lacks permanence. Where Speed offers verbal dexterity, the rustic clown Lance comically exploits the inability of language to signify objects. In his first speech in the play, Lance performs a comic monologue that foregrounds the challenges of matching words, objects, and bodies. He recounts the scene of his departure from home to serve Proteus:

> Why, my grandam, having no eyes, look you, wept herself blind at my parting. Nay, I'll show you the manner of it. This shoe is my father. No, this left shoe is my father. No, no, this left shoe is my mother. Nay, that cannot be so neither. Yes, it is so, it is so: it hath the worser sole. This shoe, with the hole in it, is my mother, and this my father. A vengeance on't— there 'tis. Now, sir, this staff is my sister; for, look you, she is as white as a lily and as small as a wand. This hat is Nan, our maid. I am the dog. No, the dog is himself, and I am the dog. O, the dog is me, and I am myself. [...] Now the dog all this while sheds not a tear nor speaks a word; but see how I lay the dust with my tears. (2.3.11–20, 27–9)

Crab's inability to weep alongside Lance's family members generates a series of mimetic failures, including his inability to represent the story through words or objects, in part because of the failure of objects at hand to stand in for their intended referents. Not only does Lance fail to find the proper material vehicles for his story, the story itself details the failure of his family and maid to achieve the proper affective state in relation to his departure from them.[64] The clown's dog neither speaks nor cries—he is an ambivalent signifier of nothing. Like other self-reflective moments in Shakespeare, this one sheds light on the larger structure of theatrical representation and its dependence on language to produce action and character, a convention that Lance mocks in his comic monologue. Lance's pantomime also literalizes his metaphorical association of dog and stone. But his larger inability to align words with their objects also speaks to broader problems of theatrical mimesis.

At times in the play, Speed repeats Crab's problems of communication in the tradition of the archetypal servant in comedies by Terence and Plautus who mixes up letters in the process of delivering them.[65] Speed is a version of the Shakespearean fool whose witty wordplay confutes ordinary syntax and submits language to such overdetermined fluidity that it can no longer communicate. In the court of Milan, Speed draws attention to a non-verbal order of representation in the "special marks" that will tell him if Valentine is in love with Silvia (2.1.16–29), while Lance, for his part, asks "Can nothing speak?" of a dejected Valentine who has referred to himself as "nothing" following his banishment from Milan (3.1.197–8). Likewise, when the anxious and buffoonish lover Thurio asks Proteus how Silvia likes his "discourse," the disguised Julia notes that she probably likes it better when he holds his "peace" (5.2.17–18). And when Proteus bids farewell to Julia, who is silent, he confirms his sense that language cannot properly channel truth: "What, gone without a word? / Ay, so true love should do. It cannot speak, / For truth hath better deeds than words to grace it" (2.2.16–18).

In all of these examples, the "nature" of love entails problems of verbal and visual representation that the female protagonists of the comedy also invoke. In her opening lines, for example, Julia argues that men do not love if they do not "show" their love (1.2.31), while in a later scene she calls the false words and tears of Proteus "pure messengers sent from his heart" (2.7.77). Silvia also refuses to read a letter from Proteus because she believes it to be "stuffed with protestation" and "new found oaths" that he will "break / As easily as she can tear the paper" (4.4.124–6). Before she learns that Proteus will leave her for the court of Milan, Julia rips up a love letter from Proteus as a way to signal her dislike of him, but she quickly repents and berates her own "hateful hands" for tearing such "loving words" and for casting her lover's name "against the bruising stones." She tries to reassemble the physical letter itself as a kind of compensation for her violent act of destruction:

> And here is writ "Love-wounded Proteus."
> Poor wounded name, my bosom as a bed
> Shall lodge thee till thy wound be thoroughly healed;
> And thus I search it with a sovereign kiss.
> But twice or thrice was "Proteus" written down.
> Be calm, good wind: blow not a word away
> Till I have found each letter in the letter
> …
> Lo, here in one line is his name twice writ:
> "Poor forlorn Proteus, passionate Proteus,
> To the sweet Julia"—that I'll tear away;
> And yet I will not, sith so prettily
> He couples it to his complaining names.
> Thus will I fold them one on another:
> Now kiss, embrace, contend, do what you will. (1.2.114–20, 124–30)

Julia's efforts to reassemble the torn letter imagine the ink-stained paper as a body that can be reanimated, and her attempted reconstruction of "each letter in the letter" playfully invokes the Protestant dream of *sola scriptura*. Yet the act of reassembling the letter also gives Julia a renewed sense of her attachment to Proteus. She urges the words and folds of paper she presses together to "kiss, embrace, contend, do what you will." The comic effect of the scene demonstrates that this effort is wholly inadequate. The reconstructed letter cannot embody the lover and Proteus departs for Milan in the very next scene, leaving Julia speechless.

Shakespeare draws inspiration for the central question of love's expressibility from his source text: Jorge de Montemayor's *Diana* (1559), which Spenser and Sidney had earlier adapted into poetry and fiction.[66] The central plot of the *Diana* concerns the question of whether, as Don Felix says in a letter to Celia, "true love, without dissimulation, doth not suffer any thing to be hid."[67] Mixing song, prose, and verse across hundreds of pages, all characters have ample opportunity to express their emotional states, and the rival forms that such emotions take in the text suggest a tension in the modes of representing true love. Montemayor's romance invites its readers to examine the relative efficacy of different mediums of erotic representation without privileging one medium over another. Indeterminacy results, both in the final ambivalence in Diana and Sireno's love and in the author's failure to complete a promised sequel.

Shakespeare heightens this ambivalence in *Two Gentlemen of Verona* by emphasizing the disparity between emotion and vehicles in theatrical bodies, images, and words. Montemayor's Felix becomes Shakespeare's Proteus, an emblem of the changeability of romantic desire who also plays a fallen hero redeemed by female love.[68] Where Montemayor focuses on a single love letter from Felix/Proteus to Felissimus/Julia that remains unread, Shakespeare multiples the number of letters in his play to complicate the problem of expressing passion.[69] In act 2, for example, Silvia asks Valentine to write a letter of love to a secret friend, then rejects the love letter that Valentine has written to her, before Valentine offers to write a new letter that Silvia does not promise to read. Speed promptly identifies the "jest" of the exchange: Silvia has, in essence, urged Valentine to write letters to a rival suitor:

> For often you have writ to her, and she, in modesty,
> Or else for want of idle time, could not again reply;
> Or fearing else some messenger, that might her mind discover,
> Herself hath taught her love himself to write unto her lover! (2.1.148–51)

Here and elsewhere in the play, letters fail to express romantic desire accurately, even when properly delivered to their addressee. Materialized language reveals itself to be inadequate, yet the play does not necessarily find resolution in non-verbal alternatives, as its language of idolatry suggests.

"Your Picture for My Love"

Where the opening acts of *Two Gentlemen* mainly concern verbal and written communication, the action at the court of Milan beginning in act 2 foregrounds images.[70] As Proteus arrives in Milan to join Valentine, he quickly forgets his verbal "oaths" to Julia and describes his sudden desire for Silvia in visual terms: "'Tis but her picture that I have yet beheld, / And that hath dazzled my reason's light" (2.4.205–6). The lovesick Valentine, suffering "bitter fasts" and "penitential groans" at the altar of Love (2.4.127), praises the divinity of Silvia:

PROTEUS
Enough! I read your fortune in your eye.
Was this the idol that you worship so?
VALENTINE
Even she; and is she not a heavenly saint?
PROTEUS
No, but she is an earthly paragon.
VALENTINE
Call her divine.
PROTEUS
I will not flatter her.
VALENTINE
O, flatter me; for love delights in praises. (2.4.139–44)

To call Silvia an "idol" is to name her as an unworthy object of love, a mere source of flattery who is only a pale imitation of divine attraction. Ironically, however, Proteus anticipates his own infatuation with Silvia, which is revealed as idolatrous in a different way. For Proteus, love is idolatry precisely because it is shifting and inconstant, illustrated by his sudden pursuit of Silvia. Shakespeare highlights the changeability of love in terms of idolatrous images, as Proteus describes his old love for Julia as "thawed… / like a waxen image 'gainst a fire" (2.4.196–7).

Silvia's status as an object of desire in Milan inspires increasingly idolatrous forms of attraction from both Valentine and Proteus. The disguised Julia arrives at the court to witness the most explicit act of erotic idolatry, as she observes her erstwhile lover woo Silvia. After trying and failing to win Silvia by singing the amatory poem "Who is Silvia?" under her window (4.2.37–52), Proteus attempts a different medium of amorous pursuit:

PROTEUS
Madam, if your heart be so obdurate,
Vouchsafe me yet your picture for my love,
The picture that is hanging in your chamber.
To that I'll speak, to that I'll sigh and weep;
For since the substance of your perfect self.
Is else devoted, I am but a shadow,
And to your shadow will I make true love.
JULIA (*aside*)
If 'twere a substance, you would sure deceive it
And make it but a shadow, as I am.

SILVIA
I am very loath to be your idol, sir.
But since your falsehood shall become you well
To worship shadows and adore false shapes,
Send to me in the morning, and I'll send it,
And so, good rest. (4.2.113–24)

The painting offers another "noteworthy object" anticipated by Proteus in the opening lines of the play, an emblem of the logic of substitution that manifests itself in the central love plot. Failing to woo Silvia through spoken and written language, Proteus worships instead the painted "shadow" of her portrait as a substitute for her living presence. (Julia's aside suggests that Silvia herself would become a "shadow" or spurned lover even if Proteus were successful in his suit.) The dialogue combines the Platonic dyad of substance and shadow with a Judeo-Christian discourse of idolatry in expressing the logic of substitution at the erotic heart of romance.[71] At the same time, Silvia undercuts this idolatry by noting that she is "very loath" to serve as an "idol."[72]

Julia's verbal portrait of Silvia's painting elevates verbal language over images, at least temporarily, though that language also attempts to incorporate the erotic energy of those images:

Come, shadow, come and take this shadow up,
For 'tis thy rival. O thou senseless form,
Thou shalt be worshipped, kissed, loved and adored!
And, were there sense in his idolatry,
My substance should be statue in thy stead.
I'll use thee kindly for thy mistress' sake,
That used me so; or else, by Jove I vow,
I should have scratched out your unseeing eyes
To make my master out of love with thee! (4.4.192–200)

Here the idolatrous desire of Proteus for Silvia contaminates Julia's own expression of desire for Proteus. Even as she threatens to destroy the image in a fit of iconoclasm, Julia hopes that her "substance" might be transformed into the "statue" or painting that Proteus will "worship."

The painted picture of Silvia both improves her image ("the painter flattered her a little" [4.4.182]) and reveals minor differences between Silvia and Julia—one has auburn hair and one golden hair, though either could wear a "colored periwig" that would allow one to impersonate the other (4.4.186). Mirroring the process that a boy actor in the original production would have taken in order to play the role, Julia must change her shape, appearance, and gender in order to follow Proteus and thus to remain true to her lover.[73] Shakespeare in fact makes a direct link between Julia's transformation and theatricality by alluding to her performance in a Whitsun pastoral in which she played "the woman's part" (4.4.155), wearing a gown while playing the role of Ariadne, "passioning / For Theseus' perjury and unjust flight" (4.4.162–3). This Ovidian metamorphosis links her to Proteus—but where his outside remains constant while his heart shifts allegiance, her heart remains true while her appearance changes.

As Jonathan Bate notes, Julia's portrayal of Ariadne alludes to book 10 of Ovid's *Heroides* (c. 15 BCE) where she laments her absence from Theseus and her exile from her home.[74] Her despair is such that she wishes herself dead like her brother, who was slain by Theseus himself. She thus represents a woman "buried by the treason of a perjured mate" (10.76), in Ovid's words—an appropriate analogue of Julia's sadness in response to her lover's betrayal. But Ariadne also appears in book 8 of Ovid's *Metamorphoses*, where she helps Theseus slay the Minotaur by giving him a thread that he uses to retrace his steps in the maze designed by Daedalus. The minotaur trapped in the maze on the order of Theseus is the product of adulterous bestiality between Pasiphaë, wife of Theseus, and a bull. The help that Ariadne renders Theseus anticipates the aid that Scylla gives to the invading King Minos, enemy of her father Nisus, in the opening tale of book 8. In fact, as she falls in love with Minos before betraying her father by cutting off his magical "purple lock," Scylla engages in a version of erotic idolatry by imagining Minos as the living embodiment of Apollo (8.9). Wishing herself transformed into the spear and reins that Minos fingers, she betrays her father and her country by delivering the patriarchal lock into the hands of his enemy. Ariadne in book 8 is thus a figure of considerable pathos whom Ovid associates with the power of representation to transform the material world.

Julia's reference to Whitsunday also highlights the play's own status as a work for the theater, linking Shakespeare's comedy with earlier modes of drama such as those we examined in Chap. 2. Whitsun pastorals were increasingly targeted by reformers in the sixteenth century because of their perceived secular and pagan focus.[75] Julia does not necessarily contradict the claim of antitheatricalists that the theater is a source of idolatry. Yet she also reveals how Shakespeare reimagines the status of performative bodies in relation to images and words. Her soliloquy anticipates the more fully embodied "presence" of Silvia in act 5.

The "Presence" of Sylvia

Proteus senses the danger and ineffectiveness of his idolatrous passion for Silvia, as he confesses in act 4 that she is "too fair, too true, too holy, / To be corrupted" with gifts (4.2.5–6). His idolatrous love also makes him "false" to Valentine, suggesting that idolatry governs both homosocial and heterosexual bonds in the play.[76] In the forest scene of act 5, Silvia urges Proteus to return to Julia using the language of textual inscription and interpretation:

> Read over Julia's heart, thy first best love,
> For whose dear sake thou didst then rend thy faith
> Into a thousand oaths, and all those oaths
> Descended into perjury to love me!
> Thou hast no faith left now unless thou'dst two,
> And that's far worse than none. Better have none
> Than plural faith, which is too much by one,
> Thou counterfeit to thy true friend. (5.4.46–53)

Silvia argues that the idolatrous affection of Proteus produces a dangerously "plural faith." Her argument recalls the argument of Shakespeare's sonnet 105 ("Let not my love be called idolatry"), which insists that underneath the diversity of expressions of love offered by the sonnet sequence as a whole there is a unifying truth that the beloved is "fair, kind, and true" (lns. 10, 11, 14). Silvia urges Proteus to "read over" her heart in order to renew his original bond to Julia and cease being a "counterfeit" to his friend Valentine.

But Proteus does not listen to the words of his presumed beloved. In his attempted rape of Silvia, Proteus reveals himself to be an absolute "counterfeit" both to Valentine and to the norms of courtly love he claims to uphold, not to mention a direct threat to Silvia herself. The scene also suggests that his threat of sexual violence is a logical extension of his idolatrous attraction to Silvia. A line spoken by Valentine at the opening of act five, scene four—"Repair me with thy presence, Silvia" (5.4.11)—illustrates this logic. The embodied presence of Silvia in the scene, as opposed to the textual or visual representation of her in letters and images earlier in the play, has redemptive power. Julia emphasizes this power of Silvia a few lines later when she predicts her own unhappiness as Proteus "approacheth [Silvia's] presence" (5.4.32). Proteus attempts to possess this "presence" by force, stopping only because of the intervention of Valentine. This expression of shame and guilt by Proteus, together with his request for forgiveness from Valentine, prods Proteus to accept the "repentance" of his friend (5.4.80) and to offer what was his own "in Silvia" to Proteus (5.4.84).

Following the logic of idolatry that Shakespeare employs throughout *Two Gentleman*, Silvia's true presence overcomes the false idolatry of erotic obsession that has consumed Proteus. Her lively body provides an antidote to what the art historian David Freedberg calls the "cold materialism" of idolatry and asserts the power of the theater to generate a reformed "presence" on stage.[77] This "presence" rewrites the empty signifiers of idolatry that have appeared in earlier scenes in the form of painted images, broken vows, and misdelivered letters. This physical "presence" of Silvia in the play enables the reformation of the male lovers away from a Catholic form of idolatrous infatuation and toward a more restrained mode of adoration directed toward marriage. The restorative power of Silvia offers a panacea to the failed conjunctions of words, objects, and images in earlier scenes. Of course, the fact that the "body" beneath this presence is that of a boy playing a female adds an imoprtant layer of this reading.[78] Such "presence" is artificial and constructed, but it also blurs the distinction between idolatrous and non-idolatrous representation. Shakespeare's play must, in this sense, represent idolatry in order to reform it, and the audience participates in this process.

It is also noteworthy that this reparative "presence" of Silvia occupies a forest—a green space located somewhere between Mantua and Milan where outlaws modeled after Robin Hood have been exiled from the court. The rough inhabitants who have been removed from polite society for both petty and serious crimes invite the disgraced Valentine to be their "captain" (4.1.64). This is the same forest where Silvia flees with the help of the elder courtier Eglamour rather than be forced by her father to marry Thurio. Less a pastoral retreat than a space of what Giorgio Agamben calls "bare life," the forest of the play is antithetical to the court of Milan in every way.[79] Valentine calls the "unfrequented woods" a "shadowy desert" that he can "better brook than flourishing peopled towns" (5.4.2–3). No longer forced to play the language games of the court in order to achieve favors, he remains "alone, unseen of any" (5.4.4). It is only in such a barren and silent space that Valentine can become authentically worthy of Silvia's love. Indeed, the play reforms histrionic excess of idolatrous eroticism through enacting this coupling, channeling misguided passion into heterosexual marriage.

In its efforts to neutralize idolatry, moreover, the play transforms the potentially transgressive forms of sexuality it stages into new and "godly" domestic structures.[80] The central language of idolatry codes this domestication in terms of a Protestant tempering of Catholic forms of passion. Where Catholic authors encouraged emotional affinity with "Mother Jesus," Reformation portraits of faith were couched in more conventionally masculine language of active might and energetic faith.[81] The silencing of Silvia and Julia's female "presence" in Shakespeare's comedy reproduces this process for a theatrical audience, elevating male penance and the rational distribution of sexual partners over female agency. This gesture upholds a Protestant ideal of patriarchal marriage, yet Silvia's redemptive presence comes at the cost of the erasure of her own subjectivity through marriage. It also emerges, as we have seen, as part of Shakespeare's interest in the order of nature as a source of theatrical representation. Moving between objects, language, and images as vehicles of representation, Shakespeare's comedy depends on early modern languages of idolatry in order to locate, ultimately, authentic forms of attraction exclusively in the natural body. This comedy understands that body as an artificial construction of language, costume, gestures, and images—a body that must produce idolatrous desires in order to make them present for the audience, even if the play ultimately seeks to purge the charge of idolatry against the early modern theater as a cultural institution.

Notes

1. Bacon, *Novum Organum*, ed. Jardine and Silverthorne, 47.
2. On this point, see John C. Briggs, *Francis Bacon and the Rhetoric of Nature*, 189–201.
3. Bacon, *The Advancement of Learning*, ed. Kiernan, 22, 121. On Bacon's use of and interest in theories of rhetoric, see Brian Vickers, *Francis Bacon and Renaissance Prose*; Lisa Jardine, *Francis Bacon: Discovery and the Art of Discourse*; and Briggs, *Francis Bacon and the Rhetoric of Nature*.
4. Bacon, *The Advancement of Learning*, ed. Kiernan, 127.
5. Ibid., 74,
6. Bacon, *The Advancement of Learning*, ed. Kiernan, 73.
7. Bacon, *The Advancement and Proficience of Learning* (1640), ed. Watts, 107. The passage itself may have been added by Watts.
8. Bacon, *The Advancement of Learning*, ed. Kiernan, 23.
9. John Webster, *Academiarum examen*, 28.

10. Bacon, *Novum Organum,* ed. Jardine and Silverthorne, 53. For a discussion of this passage in the context of a larger study of the influence of religious thought and tradition on the works of Bacon, see Steven Matthews, *Theology and Science in the Thought of Francis Bacon,* 110.
11. This passage comes from Bacon's "Plan of the Work" prefacing *Novum Organum*, ed. Jardine and Silverthorne, 11.
12. Bacon, *Novum Organum,* ed. Jardine and Silverthorne, 37.
13. Bacon, *The Advancement of Learning,* ed. Kiernan, 154. See also Peter Harrison, "The 'Book of Nature' and Early Modern Science," 8.
14. Bacon, *The Advancement of Learning*, ed. Kiernan, 106.
15. Bacon, *Novum Organum,* ed. Jardine and Silverthorne, 61.
16. See the copy of this text in *The Works of Francis Bacon,* ed. Spedding, Ellis, and Heath, 10:408.
17. This passage appears in the 1640 translation of the Latin expansion of the 1605 text by Bacon, *The Advancement and Proficience of Learning,* trans. Watts, 80. See also *Works,* ed. Spedding et al., 8:410–11.
18. Bacon, *The Advancement of Learning*, ed. Kiernan, 116.
19. Bacon, *Novum Organum,* ed. Jardine and Silverthorne, 40–2.
20. Thomas Browne, *Pseudodoxia epidemica* (1634), ed. Killeen, 117. All quotations from this text are from this edition, hereafter cited parenthetically in the body of the text by page number.
21. On hieroglyphs more broadly, see Erik Iversen, *The Myth of Egypt and its Hieroglyphics in European Tradition,* and Rudolf Wittkower, "Hieroglyphs in the Early Renaissance." On Browne's relationship to hieroglyphics, see Gordon K. Chalmers, "Hieroglyphs and Sir Thomas Browne," and Thomas C. Singer, "Sir Thomas Browne's 'Emphaticall decussation, or fundamentall figure.'"
22. As Singer suggests, the language of hieroglyphics is "natural because it is written into nature, because nature is its script, and because it is not conventional—the essential connection between signs and referents that makes its symbols substantial are guaranteed by their having been written by God Himself." See "Geometrical Hieroglyphics," 98.
23. See Watson, *Back to Nature*, 38.
24. "Principio id est narrare; secundo id est decipere; tertio sit voluntatem propriam habere seu operare in opere; quarto sit adorare ydola." The manuscript, housed in the Bibliotheka Uniwersytecka (IQ 158) in Poland, was posted to the REED Internet Bulletin Board by Stephen James Killings. Andrew Gurr cites it and discusses it briefly in *Shakespeare's Workplace*, 151.
25. Julia Lupton remarks that "the play's rationalizing deflation of its carefully staged mystery definitively undercuts the Catholic iconography the scene so powerfully evokes, enacting the movement from the Church to its Reform" in *Afterlives of the Saints*, 216. Barbara Newman shares this view in *Medieval Crossover*, 7. Emma Maggie Solberg elaborates on these views in light of Catholic traditions of articulable statues in "Mirror Images."

26. Several scholars have located important religious themes in the play. Maurice Hunt offers perhaps the fullest account in "Catholicism, Protestant Reformation, and *The Two Gentlemen of Verona*," which discusses idolatry briefly. He also addresses the play in "*The Two Gentlemen of Verona* and the Paradox of Salvation." René Girard explores some central religious metaphors in the play in "Love Delights in Praises." See also Michael O'Connell's brief discussion of the play in *The Idolatrous Eye*, 128.
27. On Hildegard see Bernard McGinn, "Hildegard of Bingen as Visionary and Exegete." On Fiore, see B. Hirsch-Reich, *The Figurae of Joachim of Fiore*. McGinn suggests that both Hildegard and Fiore were interested in "showing the action of the triune God in the history of salvation" rather than trying to "manifest how humans during this life can share in the hidden inner life" of the Holy Trinity, something he ascribes to Meister Echart and his pupil Henry Suso. See "Theologians as Trinitarian Iconographers," in *The Mind's Eye*, 195.
28. Jean-Luc Marion, *The Crossing of the Visible*, 98, n. 12.
29. Martin Luther, "Against the Heavenly Prophets," in *Works* 40:99. Cited in Watson, *Back to Nature*, 39.
30. Martin Luther, *Invocavit Sermons*, in *Martin Luther's Basic Theological Writings*, 298.
31. For a fuller account of the debate between Luther and Karlstadt regarding idolatry, see Carlos Eire, *War Against the Idols*, 55–74.
32. William Tyndale, *The Obedience of a Christian Man* (1528), sig. L8v.
33. Ibid., sig. P.
34. See, e.g., Milner, *The Senses and the English Reformation*, esp. chapters 7 and 8.
35. Henry Holland, *The History of Adam, or the Four-Fold State of Man* (1606), fol. 42v.
36. Queen Elizabeth I, *The Second Tome of Homilies* (1563), 42.
37. Lawrence Stone, "Literacy and Education in England, 1640–1900," 78.
38. Patrick Collinson, *From Iconoclasm to Iconophobia*.
39. See, e.g., Shannon Gayk, *Image, Text, and Religious Reform in Fifteenth-Century England*, 1–14.
40. See, e.g., Tessa Watt, *Cheap Print and Popular Piety, 1550–1640*; and Tara Hamling, *Decorating the 'Godly' Household: Religious Art in Post-Reformation Britain*. As David J. Davis puts it, "at least some Protestants, even in the later sixteenth century, did not despise all religious imagery, even as they destroyed some images." See *Seeing Faith, Printing Pictures*, 2.
41. Elizabeth Truax locates a tradition of moralistic visual decoration in Protestant England in wall paintings in inns and taverns that derive from emblem books. See "Emblematic Pictures for the Less Privileged in Shakespeare's England."

42. These quotations regarding the Rood of Grace come from G.H. Cook, *Letters to Cromwell and Others on the Suppression of the Monasteries*, 144.
43. See George Cornelius Gorham, *Gleanings of a Few Scattered Ears*, 17–19. Also see John Phillips, *The Reformation of Images*, 73–4.
44. On the spectacular destruction of the Boxley Rood of Grace in relation to theatrical performance, see Leanne Groeneveld, "Lyke unto a Lyvelye Thyng."
45. *Calvin's Commentaries on The Acts of the Apostles*, 2:118–9. Jennifer Waldron cites this passage to highlight God's enduring presence in nature and his power in the world, drawing the distinction between "dead" images of idolatry and "lively" images of God. See her *Reformations of the Body*, 49.
46. Zwingli, *Commentary on True and False Religion*, cited in Eire, *War Against the Idols*, 85. Calvin, *Institutes of the Christian Religion* (I.v.9), trans. Beveridge, 1.15.3. See also Susan Schreiner, *The Theater of His Glory*, 95.
47. Calvin, *The Institution of Christian Religion*, trans. Norton, sig. A5. All citations to Calvin are to this edition, hereafter cited parenthetically in the text by signature.
48. Calvin published the *Institutes* originally in Latin in 1536 before helping with a translation into French. An English translation appeared in London in 1561 as the *Institution of Christian Religion.* A Latin edition was printed in 1576. For the Latin version of the *pulcherrimo theatro* passage, see Calvin, *Institutio Christianae religionis*, 70.
49. According to Margaret Miles, Calvin relieves the body of sin by locating both the "spirit" and the "flesh" in the soul. See "Theology, Anthropology, and the Human Body in Calvin's *Institutes of the Christian Religion*," esp. 311–17. Waldron likewise argues in *Reformations of the Body* that the body in Calvin is "a divinely appointed medium through which [all Christians] could understand and participate with God during their lives on earth" (59). For her, "it was in and through their bodies that worshippers could confirm their link to the sacred" (73).
50. On the ironic tension between the plenitude of divine sources of knowledge and the inability of sinful humans to understand it, see Calloway, *Literature and Natural Theology in Early Modern England*, 32–5.
51. As Robert N. Watson has argued, the Protestant position of consubstantiation explodes the playfulness and paradox of the High Church position that understood Christ's body as both present and absent at the same time, replacing such duality with a narrower choice between communion as memorial reconstruction of Christ's memory or as his literal body and blood. See *Back to Nature*, 250

52. James Simpson notes that Henry VIII attempted to "contain" the "revolutionary forces" caused by his break with Rome through a series of proclamations that attempted to define ceremonial "signs" as forms of remembrance and not presence. "The proclamation made the present an area of remembrance alone, rather than a field through which the energies of the past continued to flow." See *Reform and Cultural Revolution*, 384.
53. Joseph Leo Koerner, *The Reformation of the Image*, 311. See also Watson, *Back to Nature*, 46.
54. On Luther's theory of the eucharist, see Thomas J. Davis, "The Truth of the Divine Words" and Thomas Osborne, "Faith, Philosophy, and the Nominalist Background to Luther's Defense of the Real Presence," 63–82.
55. On Luther's claims about the spirituality of all vocations, see Stephen Gaukroger, *The Emergence of a Scientific Culture*, 104.
56. Cited in Marguerite A. Tassi, *The Scandal of Images*, 40.
57. Philip Stubbes, *Anatomy of Abuses*, sig. L5v. Reprinted in Pollard, ed., 118.
58. William Rankins, *A Mirror for Monsters*, sig D1v and 24. Reprinted in Pollard, ed., 132–133.
59. Stephen Gosson, *Plays Confuted in Five Actions*. Reprinted in Pollard, ed., 98–100.
60. "Eutheo" [Anthony Munday?], *A Second and Third Blast of Retreat from Plays and Theaters*, 69.
61. William Shakespeare, *Mr. William Shakespeare's Comedies, Histories, & Tragedies*, sig. A4v.
62. As Freccero remarks: "My point is simply that idolatry, however repugnant to an Augustinian moralist, is at the linguistic level the essence of poetic autonomy. Because language and desire are indistinguishable in a literary text, we may say that by accusing his persona of an idolatrous passion Petrarch was affirming his own autonomy as a poetic creator" in the *Canzoniere*. See "The Fig and the Laurel Tree," 40.
63. Richard Wilson observes that the early modern theater was a "special medium for articulating—or disarticulating—semiotic problems of presence and absence." See *Christopher Marlowe*, 45.
64. On the "unyielding materiality" of Lance's shoes in relation to the protean language of the play, see Malcolm Evans, "Deconstructing Shakespeare's Comedies," 70–1.
65. On Speed and Lance as forces of satirical critique of courtly values in the play who derive from popular traditions of humor, see Robert Weimann, "Laughing With the Audience."
66. In fact, Montemayor's work established a vogue for pastoral romance in the second half of the sixteenth century. See Judith M. Kennedy's introduction to her *Critical Edition* of Montemayor, xxi–lvii.
67. Ibid., 96.

68. Tom Shippey traces the literary origins of Proteus and links him to the theme of "substance without fixed shape" in "The Protean Allegory of Matter in *The Two Gentlemen of Verona*," 26. William O. Scott identifies the Ovidian and Spenserian origins of Proteus as a variable lover in need of "fixing" by Julia in "Proteus in Spenser and Shakespeare."
69. On Shakespeare's use of letters in the play, see John A. Gwinn, "The Letter Device in the First Act of *Two Gentlemen of Verona*," and Frederick Kiefer, "Love Letters in *The Two Gentlemen of Verona*." Jonathan Goldberg also reads the letters as origins of character in the play in *Voice Terminal Echo*, 68–81.
70. Catherine Bates regards *Two Gentlemen of Verona* as one of Shakespeare's "most pointedly epistolary plays" but does not attend to the juxtaposition of images and spoken words in her chapter on "Love and Courtship" in *The Cambridge Companion to Shakespearean Comedy*,. 112.
71. Shippey reads Proteus and Julia as metaphorical shadows or "wasted forms of their true shelves" in "The Protean Allegory," 27.
72. Hunt, "Catholicism, Protestant Reformation," 6.
73. On the early modern practice of using boy actors in early modern drama, see Evelyn Tribble, "Pretty and Apt: Boy Actors, Skill, and Embodiment."
74. Jonathan Bate, *Shakespeare and Ovid*, 120.
75. See, e.g., Glynn Wickham, *The Medieval Theater*, 133. Wickham understands Shakespeare's references to "Whitsun" pastorals and Pentecost plays in *Two Gentlemen of Verona*, *Love's Labour's Lost*, *A Midsummer Night's Dream*, and *The Winter's Tale* as acts of both mockery and nostalgia (148).
76. Masten, "*Two Gentlemen of Verona*," 273.
77. David Freedberg, *The Power of Images*, 379.
78. For a broad and compelling account of the enduring practice of using boy actors in early modern theater, see Stephen Orgel, *Impersonations*.
79. Although defined mostly in terms of twentieth-century politics, "bare life" is a condition in which people must confront basic challenges of survival and reproduction of their life without the benefit of state protection. See Giorgio Agamben, *Homo Sacer*.
80. Caroline Walker Bynum, *Jesus as Mother*, 185.
81. See Merry Wiesner, "Luther and Women," 131–2.

CHAPTER 5

Natural Perspectives in the Henriad

As we saw in the introduction, plays for Hamlet are mirrors that aim to "show the very age and body of the time his form and pressure" while also embodying and testing forms of nature. But when Hamlet meets the foppish courtier Osric in act 5, Shakespeare mirrors Hamlet's earlier mirror back at him. Osric's outlandish praise of Laertes as "an absolute gentleman full of most excellent differences" (5.2.91–2) intends to move Hamlet to accept the invitation from Claudius to enter a fencing competition, but Hamlet makes a number of different points in his response:

> Sir, his [i.e., Laertes's] definement suffers no perdition in you, though I know to divide him inventorially would dozy th'artihmetic of memory, and yet but yaw neither in respect of his quick sail; but in the verity of extolment, I take him to be a soul of great article and his infusion of such dearth and rareness as, to make true diction of him, his semblable is his mirror, and who else would trace him, his umbrage, nothing more. (5.2.97–104)

Hamlet's outrageously ornate diction and syntax outdoes the lapidary speech of his ostentatious interlocutor. Osric's brief response, "Your lordship speaks most infallibly of him" (5.2.105), suggests that he, like many in the play's audience, struggles to understand Hamlet's basic meaning. Looking more carefully at Hamlet's words, however, we see a complex relation between images, words, and objects that reflects the hermeneutics

A. Kitch, *Shakespeare's Theater of Nature*, Palgrave Studies in Literature, Science and Medicine,
https://doi.org/10.1007/978-3-031-78082-0_5

of Shakespeare's theater. In his rhetorically elaborate speech, Hamlet argues that Laertes so perfectly embodies the ideal courtier that those who try to imitate him can only imitate his shadow ("umbrage") or outline, since his essence is unique.

This exchange recalls earlier interest by Hamlet in the relation between essence and appearance. It also reaffirms Hamlet's obsession both with mirrors and modes of imitation writ large. Even in mock praise of Laertes as an ideal courtier of the kind that Hamlet himself aspires to be, the passage invokes central questions about Shakespeare's theater as a visual medium. Such problems are endemic in *Hamlet,* beginning with questions about how to see and understand the ghost. The tragedy continually stages epistemological challenges around sight, participating in what the historian Stuart Clark called a crisis in early modern visuality following the revival of classical skepticism, Protestant abhorrence of visual forms of faith, and new accounts of optics by natural philosophers.[1]

Shakespeare's visual language is so extensive that it has earned its own dictionary.[2] His plays often situate problems of vision and visuality in relation to questions of theatrical mimesis, problems that take on new dimensions in early modern playhouses that shaped acts of spectatorship in new ways. Contrary to a conventional claim that early modern audiences usually thought of going to "hear" rather than "see" a play, Gabriel Egan has unearthed evidence that visual references outnumbered allusions to hearing by a ratio of roughly 12 to 1.[3] But scholars have not fully addressed how Shakespeare's modes of visuality respond to classical, medieval, and early modern debates about vision that blend science and theology. Like optics itself before the eighteenth century, problems of vision in his plays are usually both ethical and epistemological. Experimenting with relations between distanced viewing and up-close inspection, Shakespeare joins the historical reconceptualization of hermeneutics that Timothy Reiss has described as a shift from "discursive *exchange within* the world" to a more "perspectival and incomplete process of knowing."[4] For Reiss, such a transformation entails a philosophical movement away from a world in which signs and objects are "ontologically equal" and toward a world where forms of representation become more detached from the world of things.[5] This chapter focuses in particular on Shakespeare's visual language and staging of English history, especially in his portraits of Richard II and Henry V, where he revises the "mirror of princes" genre in some of his sources to question rather than uphold the principle of divine monarchy. I suggest that these plays reveal Shakespeare's stagecraft as a dynamic mode of visuality in which perspectival viewing often challenges conventional narratives of divine power. Against the abstract theory of visual perspective, which early modern painters and masque makers employed in the practice of creating the illusion of depth, Shakespeare foregrounds a decentralized and de-divinized stage experience that embraces the individual and shifting perspectives of spectators, especially those who move around the theater during performances.[6]

How to See

Efforts to observe nature more directly in early modern Europe renewed debates about the epistemological status of vision in relation to other senses. Such efforts were also invested in a lengthy and influential tradition of divinely oriented natural philosophy. Most ancient natural philosophers supported the theory of extromission, in which the eye emits particles outward that reflect back against objects to produce images. Originally proposed by Empedocles in the fifth century BCE, this theory of vision was embraced by Plato and the Greek surgeon, physician, and rhetorician Galen, among others. Like Plato, Augustine supported extromission, but he gave it a new twist in outlining two types of vision—spiritual and everyday—that engaged in a complex dance. In everyday vision, the image that results through extromission lingers within the viewer for some time, shaping his or her soul. Citing examples from nature such as the chameleon and a pregnant woman whose baby takes on physical characteristics in response to what she sees, Augustine's theory of extromission suggests that people are shaped in fundamental ways by what they see. Human efforts to see God, by contrast, are necessarily "fleeting," since the image of God cannot be fixed or retained.

Late medieval and early modern natural philosophers increasingly turned away from extromission following the circulation of Latin translations of treatises by Alhazen (Hasan Ibn al-Haytham) in thirteenth-century Europe. A newer theory of "intromission" argued that tiny particles or *species* reach the eye externally before being delivered to the brain to be processed. Roger Bacon developed the theory in England in his *Perspectiva* (1267), which synthesized Alhazen with Greek and Roman authors such as Galen, Aristotle, and Ptolemy. According to Bacon, only God can achieve direct vision, while angels see through refraction and humans are confined to even weaker modes of vision. This argument echoes the biblical view of St. Paul that everyday vision is like trying to view the world through a glass darkly, while the vision of God achieved in heaven is face-to-face (1 Cor. 13:12). Peter of Limoges drew on the visual theories of Roger Bacon in his *Moral Treatise of the Eye* (c. 1280) to define vision in relation to Christian theology and moral philosophy. Beginning with the physical description of the eye and its method of sight, Limoges argues that the will of the observer is required for vision to function, while later chapters develop the moral and spiritual goals of vision, including an account in chapter 15 of the properties of divine sight. For Limoges, God's all-seeing eye is both spiritually healing and strengthening.[7]

In his printed encyclopedia entitled *The Pearl of Philosophy* (*Margarita philosophica*, 1496), Gregor Reisch distinguishes between *lux*—the invisible essence of light and color—and *lumen*, which produces material vision when *lux* intersects with material substances. This model of optics uses the account of the divinely fashioned light in Genesis as a model for thinking about how invisible divinity intersects with the visible world.[8] Reisch embraced and helped to popularize the concept of *species* in intromission to describe the flow of visual rays from objects into the eye.[9] Such a mixture of theology, optics, and moral philosophy defines vision not just as the physical operation of light through the eye but rather as a matter of what A. Mark Smith calls "sight"—a more fully epistemological, theological, and psychological mode of visuality in which human will determines what can be seen.[10] As opposed to light, sight transforms seeing into types of believing.

The increasing popularity of intromission in late medieval Europe generated new skepticism about the relation of the external world to the internal brain of the viewer. Theories of *species* foregrounded by Bacon and Reisch operated on a principle of imitation. The brain produced and multiplied its own images in relation to external phenomena. This is why Bacon argued that such images could be produced even when they are impossible to sense with the eye. Where extromission allowed for participation of the viewer with the forms of light that traveled from the eye to objects in the world and back again, representationalist models of vision associated most fully with Duns Scotus in the thirteenth and fourteenth centuries understood vision and external knowledge more broadly as a representation of the world.[11] It was a small step from such intromission to an understanding of the eye itself as a mode of visual technology designed like the *camera obscura*, a position given even greater strength by expanding knowledge of the anatomy of the eyeball in the sixteenth century. In his highly detailed drawings of the eye, for example, Vesalius argued that he had located the crystalline lens at its center, confirming through visual demonstration a hypothesis of Galen.[12] Like the example of Galileo's discovery of sunspots considered in Chap. 2, this effort to gain direct and unmediated knowledge of nature uncovered structures of representation within that nature. The Basel anatomist Felix Platter also employed a direct analogy between the human eye and a camera obscura in his work on the "Structure and function of the human body" in the 1580s.[13] Many natural philosophers and theologians argued, moreover, that only those with sufficient faith in God could truly see the world rather than just take it in. Representationalists such as Cusanus, however, emphasized that the human mind could never truly create something new.[14] We can only assimilate elements of divine creation in ways that benefit God.

Similar positions were taken by other early modern investigators of the order of nature, including the French natural philosopher Pierre de La Primaudaye, whose *French Academy* (*Académie Française*, 1577) Shakespeare likely used as a source text for *Love's Labors Lost*. La Primaudaye laments that "there are but few that have such eyes as are requisite for the reading of this book [of nature], remarking that good readers require both 'liberal arts' and 'humane philosophy' in order to discern the presence of God in the world."[15] Such skills for La Primaudaye extend the capacity of the divinely sculpted eye itself, which he describes in terms of "a little round Christal glass" or mirror.[16] This round eye not only peers at divinely created matter, it is itself structured like a series of mirrors that represent a "variety" of images through "figures and makings."[17] In other words, the divine eyeball is akin to a mirror that produces and receives multiple images.

La Primaudaye joined other early modern natural philosophers in taking advantage of new tools of visual technology such as the telescope or "perspective glass" to study the *ordo creationis*. As we saw in Chap. 2, such visual aids proliferated in early modern Europe thanks to innovations in design by figures such as the Dutch astronomer and inventor Christiaan Huygens, who developed new types of lenses with his brother in order to build refracting telescopes.[18] Yet most early adopters of telescopes used them to observe the earth rather than the stars. Even Galileo acknowledged their utility in this way. In a letter to the Doge of Venice written from 1609, he celebrated the telescope as a "great help in maritime and land enterprises," especially in "discovering the ships of the enemy two hours before they can be seen with natural vision."[19] In a remarkably wide-ranging manual covering topics from water-based cures to constructing fireworks in the shape of a dragon, moreover, the Englishman John Bate describes a method of calculating distances using a perspective glass and a wooden frame strung in a grid with lute strings (see Fig. 5.1), which alludes to a similar method outlined by Oronce Fine in his treatise *On the Practice of Geometry* (*De re et praxi geometrica*) first published in Paris in 1586 (see Fig. 5.2). Both represent efforts to rationalize vision by making it conform to the rules of perspective.

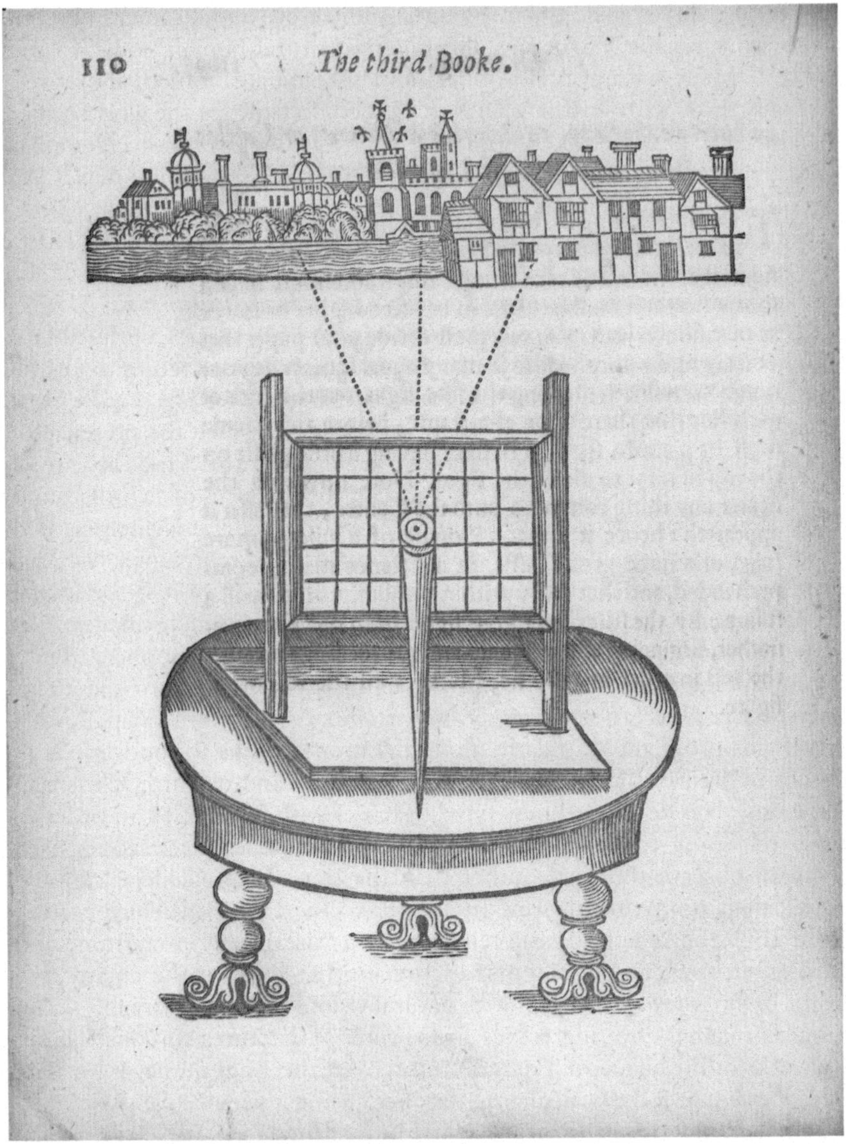

Fig. 5.1 From John Bate, *Mysteries of Nature and Art* (1634). RB 706637, The Huntington Library, San Marino, CA

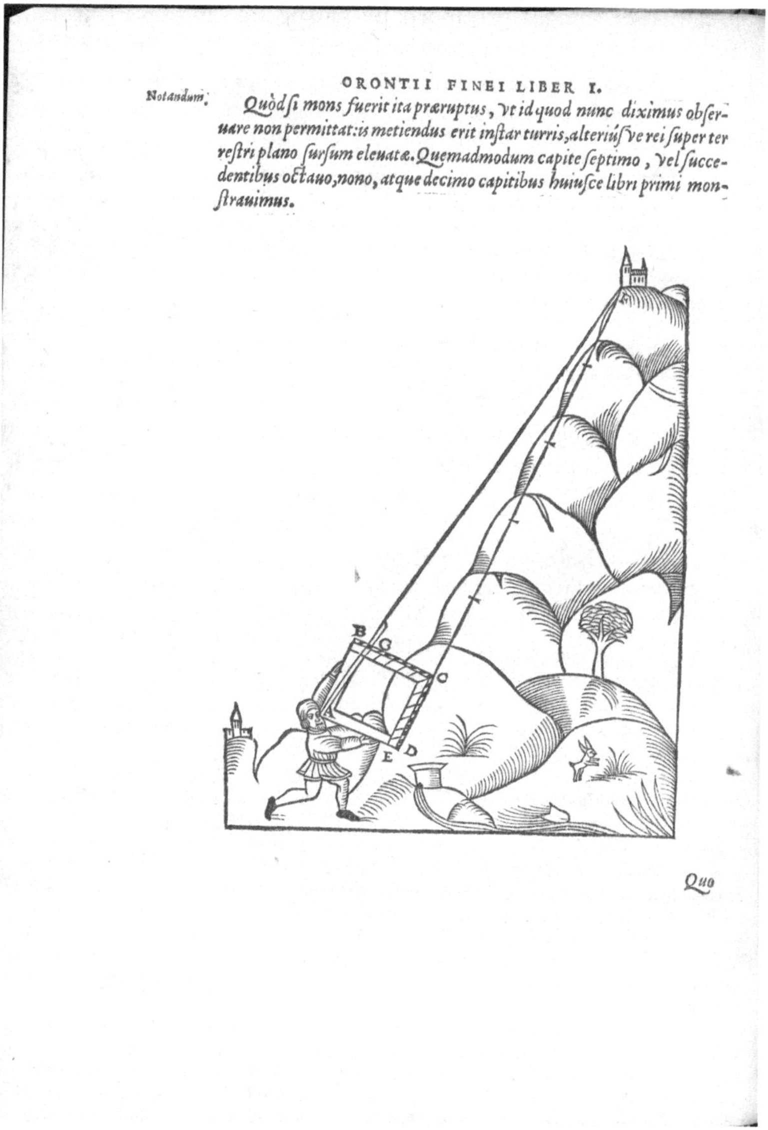

ORONTII FINEI LIBER I.

Notandum. Quòd ſi mons fuerit ita præruptus, vt id quod nunc diximus obſer-
uare non permittat: is metiendus erit inſtar turris, alteriúſve rei ſuper ter
reſtri plano ſurſum eleuatæ. Quemadmodum capite ſeptimo, vel ſucce-
dentibus octauo, nono, atque decimo capitibus huiuſce libri primi mon-
ſtrauimus.

Quo

Fig. 5.2 Oronce Fine, *On the Practice of Geometry* (Paris, 1586). RB 707857, The Huntington Library, San Marino, CA

The clarity of images produced by perspective glasses such as those pictured by Fine and Bate improved dramatically in the fifteenth century thanks to Italian crystal glass (*cristallo*) that created clearer mirrors than those typically found in northern Europe.[20] In England, the astronomer and mathematician Thomas Digges celebrated the capacity of perspective glasses to discern small objects at distances of up to seven miles.[21] Thomas Recorde in his preface to *The Pathway to Knowledge* (1551) alludes to the construction of such glasses, and Thomas Harriot was also involved in crafting lenses for one of the earliest telescopes in Europe.[22] Francis Bacon in his *New Organum* praises both telescopes and microscopes as devices that "assist the senses" in perceiving the order of nature, though he remains skeptical of the limited evidence they provide about distant worlds.[23] Bacon's elucidation of the fallibility of human senses also applies to those aided by technology. Yet proper training might alleviate such distortion. Bacon's follower Robert Boyle argued for example that accurate observation of the natural world required special knowledge:

> For the Book of Nature is to an ordinary gazer, and a naturalist, like a rare book of hieroglyphics to a child, and a philosopher: the one is sufficiently pleased with the oddness and variety of the curious pictures that adorn it; whereas the other is not only delighted with those outward objects that gratify his sense, but receives a much higher satisfaction, in admiring the knowledge of the author, and in finding out and enriching himself with those abstruse and veiled truths dexterously hinted in them.[24]

For Boyle and other members of the Royal Society, the improved vision made possible by these new visual technologies helped scientists to access "abstruse and veiled truths" hidden in the Book of Nature. As we saw in Chap. 2, celebrated discoveries in the heavens by Kepler and Galileo, among others, made with the help of telescopes popularized links between perspective glasses and astronomy, yet their use for earthly observation remained dominant.[25]

Francis Bacon was especially interested in the problematic deceptiveness of vision as an obstacle to new forms of scientific knowledge. He condemned false and dogmatic modes of received wisdom produced in "unreal and scenic fashion," for example, in describing what he called the "idols of the Theater."[26] For Bacon and likeminded natural philosophers, visual technologies such as the microscope and telescope could correct the fallibility of the human eye to produce clearer visions of the order of nature. The fullest account of such a view in early modern England was produced by the natural philosopher and polymath Robert Hooke in his *Micrographia*, which shared images made by a compound microscope that Hooke constructed himself. As Hooke notes, however, such revelation also uncovered remarkable variation in the created world. Looking carefully at insects through his microscope, for example, Hooke is surprised by how "various, and seemingly irregular are the generations or productions" he finds. Such productions cause him to

> admire the wisdom and providence of the Creator; for not only the same kind of creature may be produced from several kinds of ways, but the very same creature may produce several kinds: For, as diverse Watches may be made out of several materials, which may yet have all the same appearance, and move after the same manner [...] like watches may be wrought differing ways.[27]

Hooke remarks that the "artificial" instrument of the microscope has the potential to correct the "innate" "corruption" and "errors" of humankind, inserting the technology of vision into a narrative of redemption that Bacon also embraces.[28] The variety of forms in the fabric of nature affirms for Hooke the "wisdom and providence of the Creator," whom he compares to a "watchmaker" who can assemble working timepieces in different forms out of the same raw materials.

The analogy between a divine maker and a watchmaker was proverbial, but for Hooke it suggests a larger structure of artifice within the divine order of nature. For example, he marvels how snowflakes appear like "Artificial Figures." The more he magnifies them, he notes, the more irregular they seem. Hooke attributes such "irregularity" to the inevitable melting of the snowflakes themselves rather than to "Nature," yet his book reveals the way that the magnification of the order of nature opens up new possibilities for understanding that nature as structured artifice.[29]

Hooke argues that direct observation of the order of nature with the help of perspective glasses produces new knowledge about the "Book of Nature itself," whereas the older genre of herbals delivers nothing of the "virtues of a plant, and less of truth."[30] In fact, Hooke's microscopic study of nature signaled the demise of the herbal as a genre and even, in a certain sense, of the trope of the Book of Nature itself as a tool of inquiry. Beginning in the late seventeenth century, the trope lost its utility as a framework for justifying exploration of the order of nature.[31] For Hooke, moreover, the technology of print doubles the technology of the microscope itself in order to reproduce the order of nature in its artificial essence.

Hooke's celebration of perspective glasses countered earlier skepticism about the use of artificial technologies to enhance vision. For example, Reginald Scot, whose account of witchcraft Shakespeare employed in several plays, regarded perspective glasses as tools of the devil. In a particularly animated passage from his *Discovery of Witchcraft* (1584), Scot elaborates on the methods that Catholics can use to create false and satanic visions:

> But the wondrous devices, and miraculous sights, and conceits made and contained in glass, do far exceed all other; whereto the art perspective is very necessary. For it showeth the illusions of them, whose experiments be seen in diverse sorts of glasses; as in the hallow, the plain, the embossed, the columnary, the pyramidate or piked, the turbinal, the bunched, the round, the cornered, the inversed, the eversed, the massy, the regular, the irregular, the colored and clear glasses; for you may have glasses so made, as what image or favor soever you print in your imagination, you shall think you see the same therein. Others are so framed, as therein one may see what others do in places far distant; others, whereby you shall see men hanging in the air; others, whereby you may perceive men flying in the air; others, wherein you may see one coming, and another going; others, where one image shall seem to be one hundred, &c. There be glasses also wherein one man may see another man's image, and not his own; others, to make many similitudes; others, to make none at all. Others, contrary to the use of all glasses, make the right side turn to the right, and the left side to the left; others, that burn before and behind; others, that represent not the Images received within them, but cast them far off in the air, appearing like airy images, and by the collection of sun beams, with great force setteth fire (very far off) in every thing that may be burned.[32]

Scot's expansive catalogue of false visions generated by perspective glasses details lenses, including circular, square, dense ("massy"), cylindrical ("columnary"), and spiral or mollusk-shaped ("turbinall"), that help diabolical tricksters to mislead the eyes of their audiences. The "experiments" (i.e., experiences) of viewing the world through such mirrors do not capture the truth of nature, but rather create false if tantalizing "illusion[s]." One glass, for example, seems to produce an image of whatever the viewer imagines, while another produces an image of an absent person as the image of oneself. Such conditions of viewing for Scot are epistemologically treacherous and full of doubt, but they also reveal the degree to which the artificial science of perspective was increasingly a topic of debate in early modern Europe. Like Scot and against the technophiles of the Royal Society, Margaret Cavendish satirized perspective glasses as grossly misleading in her utopian fiction about scientific discovery, *The Blazing World* (1668).[33] Yet Cavendish shared with the members of the Royal Society she mocked a strong conviction that natural philosophy was centered on religious principles.[34]

The optical technologies that troubled Cavendish and Scot but intrigued Bacon and other natural philosophers also indicate new ways that early modern nature was understood in and through visual technologies. The glass mirrors that produced the images seen through the telescope transformed the "natural" landscape into images of two dimension that were then understood to be mimetic copies of what lay before the viewer. For some, moreover, this technologically enhanced vision became a model of how the eye itself actually worked. Human vision itself followed the arts of perspective, which employed Euclidean geometry to understand how light moves through space to create vision in the eye. But the arts of perspective first achieved prominence in early modern Europe in relation to painting and architecture, especially in the work of the celebrated Florentine engineer and architect Filippo Brunelleschi, who employed it in his design for the dome of the central cathedral in Florence, then called Santa Maria del Fiore, in the 1420s.[35] Codifying Brunelleschi's triumphs in textual form, the Florentine humanist Leon Battista Alberti adapted the principle of perspective for painters who wished to capture three-dimensional space in two dimensions. His mathematical understanding of points, lines, and planes derived from Euclid, but the optical assumptions behind his study of light and color followed the science of optics, especially the intromissionist approach of Alhazen's *Perspectiva* (c. 1021). Like Alhazen before him, Alberti describes light moving rectilinearly in extrinsic, median, and centric rays, where the centric ray enters the eye directly and the extrinsic and median rays define the visual cone that creates an illusion of depth. In Aristotelian terms, the median rays transmit the quality of things while the extrinsic rays transmit the quantities. Yet, the relation between median and extrinsic rays is liable to change as the viewer moves closer and further away from an object.[36] Such optical remapping expanded the potential range and complexity of human vision, dissolving boundaries between natural and artificial mechanisms and encouraging new varieties of artifice as models for understanding the visual order of nature.

Visual Theologies

As these examples suggest, early modern debates about optics were deeply embedded in theological principles and practices. Echoing the biblical claim by Paul in 1 Corinthians, many authors suggest, as we have seen, that only God has the capacity for direct and unmediated vision. The popular medieval and early modern genres of *imago mundi* ("image of the world") and *speculum principis* ("mirror for princes") also invested the written word with modes of insight that ultimately derive from a theological understanding of light and vision. Such genres appropriate the visual power of language in their material form as modes of textual reflection that promise to set vast amounts of knowledge before the eyes of the reader or, in the case of the *speculum principis*, to hold up a mirror to rulers in order to elevate them toward higher ideals of governance.[37]

Some modern scholars regard Alberti's theory of perspective as a materialist challenge to dominant Neoplatonic theories of art in *quattrocento* Florence, part of a broader rejection of Christianity in favor of the classical model of Euclid.[38] But Alberti's classicism was not incompatible with Christian faith, even though he frequently framed that faith in hermetic and alchemical terms. In praising "Trismegistus" (i.e., Hermes) as a "very ancient writer," for example, Alberti followed Ficino in noting that "painting and sculpture were born together with religion."[39] Other scholars interpret Alberti's theory of linear perspective in relation to the humanist pursuit of the divine within the secular—of recognizing God in nature. For Charles Carman, for example, Alberti is both rational and Catholic. Drawing on a mode of spiritual vision outlined by Nicolas Cusanus in his *Vision of God* (1453), Carman identifies in Alberti a spiritual struggle to know God, arguing that the visual pyramid denoting how one sees an object conceals on its inverse and invisible side a pyramid leading to God.[40] According to the *Oxford English Dictionary*, the word "perspective" entered the English language by way of translations of Euclid. The earliest instance of the word in the *Early English Books Online* database in 1570 derives from an edition of Euclid translated by H. Billingsley and edited by John Dee. Euclid aligns the arts of perspective with the divinity of created light—which he calls the "first of God's creatures"—and claims that it is only through understanding the art of perspective that "perfect knowledge" of the stars might be achieved.[41]

Like Euclid, Alberti assumes the divinity of nature in his theory of perspective, which he defines as an attempt to represent the world as if glimpsed through an "open window." Both regard vision as simultaneously physical and spiritual and as a product of divine creation. Painting in perspective for Alberti also follows the guide of "Nature," from which "everything is more correctly judged and measured."[42] And because nature is itself the product and manifestation of divinity, any art that employs perspective is also an art that searches out the ways of God. Subsequent painters who employed his technique of perspective harnessed the divine mathematics of Euclid in order to recreate visual illusions that had the capacity to fool the eye into seeing depth where there was none. Such deception could be regarded as diabolical, as it was for Scot, but it also revolutionized modes of visual representation.

Shakespeare's intensely visual drama follows Sidney's view that the poet should "give to the eye either some excellent perspective, or some fine Picture."[43] But Shakespeare does not follow Sidney's view that such poesy should also be morally edifying. Sidney's follower George Puttenham likewise urged authors to represent "beautiful visions" to the soul in the same way that a "glass or mirror" unifies and orders light. The mirror in this sense is more creative than reflective, providing order and unity to otherwise disorderly forces. Yet, Puttenham acknowledges that the "perspectives" provided by mirroring, as in the perspective glasses that fascinated many noblemen in the sixteenth century, could also be deforming:

> for some be false glasses and show things otherwise than they be indeed, and others right as they be indeed, neither fairer nor fouler, nor greater nor smaller. There be again of these glasses that show things exceeding fair and comely, others that show figures very monstrous and ill-favored. Even so is the fantastical part of man (if it be not disordered) a representer of the best, most comely and beautiful images or appearances of things to the soul and according to their very truth. If otherwise, then doth it breed chimeras and monsters in man's imaginations, and not only in his imaginations, but also in all his ordinary actions and life which ensues.[44]

Puttenham acknowledges the potential for mirrors to manipulate images, producing "figures very monstrous & ill-favored" that contaminate the imagination and produce not only monstrous ideas but also dangerous "actions." Puttenham also describes the historical multiplying of vision made possible by technological innovations in the period, explaining such mirroring as active rather than passive. The active construction is an essential part of the defense that Puttenham offers for imaginative writing, which has the capacity to shape the souls, minds, and actions of those who read or view it.

Although Puttenham addresses "Poesy *Dramatic*" directly in chapter 15, where he discusses those forms of imaginative writing that illuminate the shortcomings of princes, most of his examples are from poetry rather than drama. And yet, Puttenham's conception of poetry, as his writing on the creative power of the mirroring imagination suggests, is both material and constructive. As Rayna Kalas observes in her study of mirroring technologies and early modern poetry, "With the figure of the perspective glass, Puttenham is able to offer a tangible example of how the poetic image or phantasy—which is otherwise so easily dismissed as a lightheaded delusion or an ephemeral thought—could appear to be an airy figure of nothing, a mere transparency, and at the same time prove to be a technical instrument in the practice of a substantive craft."[45] In linking the art of poesy to acts of creative mirroring, both Sidney and Puttenham acknowledge the concrete and material potential of "poesy" as just such a "substantive craft."

Staging Vision

Shakespeare's intensely visual drama responded to this long tradition of philosophical and theological debate about the nature of vision while also invoking more recent concerns of Sidney and Puttenham. And he did so within the context of the nascent cultural institution of the early modern theater, which raised particular anxieties about the nature of spectacle and visuality. Puritan antitheatircalists distrusted the visual spectacle of early modern drama, yet the theater can also be seen as an agent of religious reform, as Huston Diehl and others argue. Some early modern playwrights drew on Calvin's advice to regard creation as an "open stage wherein God will have his majesty seen" while also warning against the dangers of certain images. This ethical tension in approaching objects vision defines the enterprise of early modern theater itself, as Diehl observes:

> By celebrating the created world as a divine theater wherein God manifests himself while simultaneously arousing fears that its beauty dazzles and blinds its fallen spectators, early English Protestantism contributes to the formation of a new kind of theater, one that enacts Reformation culture's struggle to master and contain the seductive appeal of the visible, imagined, and created.[46]

Though Protestants often distrusted the visual imagination—Calvin describes it as a seat of "errors" and "idolatry" for example—reformers from John Foxe onward made allowance for images that might assist in reforming Catholicism.[47] Jennifer Waldron extends Diehl's conclusions by considering how early modern theater used "lively bodies" to counteract the deadening literalism of Catholic sacramental theology. Far from serving as a vehicle of disenchantment, the early modern theater in Waldron's account depends on the fully present and sensual human body to reanimate the divine spirit within the world. Waldron joins other scholars in arguing for the reforming potential of the theater itself.[48]

The physical space of the early modern playhouse both staged and tested different modes of visuality.[49] Even with its vast open space to let in daylight, the permanent structures enclosed staged action within walls and elevated action on fixed stages to isolate visual spectacle in new ways for new audiences. Standing spectators could, for example, experience the play from multiple angles of view as part of a mobile community of playgoers. But the optics of the theater, like optics in medieval and early modern philosophy and theology, were not restricted to external phenomena. With few stage sets, Shakespeare's theater depended on moving and speaking bodies that communicated through words, song, and movement. Even as it elevated such bodies as objects of visual attention, however, these playhouses generated multiple and diverse representations.

Unlike court masques, especially in their more extravagant examples under James I, drama in popular playhouses never occupied the entire space of the viewing eye. The open-roofed playhouses did not attempt to control angles of vision in the same way as the courtly masques of Inigo Jones and Ben Jonson, who employed new stage technologies based on the principles of linear perspective. The public playhouses decentralized acts of looking by offering opportunities to view the action of moving bodies from multiple and shifting angles of vision. Moreover, actors in the theater of Shakespeare and his contemporaries returned the gaze of the spectator, multiplying the vectors and directions of vision that could occur within their open-roofed galleries. These playhouses generated artificial worlds as imitations and *re*presentations of nature, encouraging spectators to observe relations between what is seen and unseen on stage while foregrounding questions about what could be seen and known, as well as about *how* imitation itself produced knowledge about the world. Of course, historical evidence of early modern staging practices is hard to come by, even with the advantage of several archaeological discoveries of early modern playhouses.[50] Numerous maps and drawings of early modern

London depict the exterior of the buildings, but the only image of the interior of a playhouse is a sketch made by the Dutch visitor Johannes de Witt, who journeyed to London in 1596 and sent his drawing to his friend Aernout van Buchell, who made a copy (Fig. 5.3):

Fig. 5.3 Aernout van Buchell's copy of a sketch of the Swan theater by Johannes De Witt, c. 1596. (© Utrecht, University Library, MS 842. I)

De Witt drew the Swan theater with a thrust stage elevated above the ground by about five feet and twin columns that support a sloped roof connected to the "tiring house," where actors changed costumes and from which they made their entrances and exits. Paying spectators could stand and watch plays seated or standing anywhere within the approximately 270 degrees of vision surrounding the stage. This unique drawing shows spectators observing the action in a space that might also be used for balcony scenes or for the enthronement of monarchs. Atop the tiring house is a man standing next to a flag and blowing a trumpet announcing that a performance will be held on that day. In Latin text surrounding the drawing in his notebook, van Buchell records that de Witt based his image on his own eyewitness observation of the theater. Indeed, both de Witt and Buchell were skilled draughtsmen who enjoyed the popular trend at the time of making pictures of interesting objects on their travels.[51]

De Witt's drawing and accompanying labels suggest that he filtered his sketch of the Swan theater through a Vitruvian lens. The theater building appears to be circular, following the ideal theatrical structure of Vitruvius, though archaeological evidence suggests that the Swann, like other playhouses in the period, was probably polygonal. De Witt also interprets the playing space itself as the "proscenium," when the Roman theater would have used that term to designate the square frame around the stage. Likewise, the "entrance" or "ingressus" appears to refer to the space of entering the stalls rather than the entrance or exit to the stage. But De Witt struggled to identify the bare ground around the stage where spectators would normally have stood, using the vague term "planites," meaning flatness, but also writing "sive arena" ("or arena"), suggesting a Roman coliseum where enormous spectacles such as naval battles or gladiatorial combats were held.[52]

In other words, the sketch represents both the desire and failure of de Witt to capture the visual experience of the theater he attended. He mixed what he saw with his eyes with what he thought he *should* be seeing based on his humanist beliefs and practices, even if his conceptual framework was applied in a confusing and sometimes contradictory manner. Visitors to the modern Globe theater in London on the south bank of the Thames have new opportunities to experience a physical space that resembles the building pictured in the de Witt sketch. The complex and protracted process of building the modern Globe also provided new opportunities to learn how London playhouses were originally constructed.[53] For example, architects and scholars studied the complex jointing and architectural design that would have been required to raise the original structures, including the cantilevered roof and large columns supporting the overhanging "heavens" that shielded the platform stage from the elements. One of these scholars, John Orrell, concludes that the original Globe theater was a 24-sided polyhedron that was approximately 109 ft. in diameter.[54] These dimensions reveal a remarkable theory about the design of the Globe and, possibly, other early modern playhouses: they adhered to the principles of Euclidean geometry. As Andrew Gurr (who also advised the construction of the new Globe theater) observes, "if you subtend a line from the lobby wall foundations that run out from the two bays and carry it inwards to the centre of the shape, and subtend others bisecting the two gallery wall angles, all the lines come close to meeting at just over nine feet from the outer wall." The resulting shape is defined by "distinct symmetry, not just in the positioning of the stair turret, but in the size of each bay."[55] The distance between inner and outer wall of approximately twelve feet six inches seems to match that of other early modern theaters as well, particularly the Rose and the Fortune, and may also have been used to mark other dimensions in the building, including the distance from the edges of the stage to the support columns.

This possibility that the Globe and other early modern playhouses employed Euclidean geometry invites new ways of thinking about the experience of attending plays in early modern London. The dimensions of the structures and the sight lines that they produced created new viewing experiences that empowered viewers to actively engage in theatrical performances and to reflect on the modes of representation and spectatorship that such performances made possible. The physical space of the early modern public theater mingled empirical acts of observation with philosophical modes of seeing that participated in some of the questions about vision we have been tracing in this chapter. These performance spaces staged visible bodies and objects in relation to invisible and unseen presences, for example. "Discovery" spaces in the tiring house behind doors or curtains created opportunities for concealing actors, props, and sets. Additional and partially obscured spaces above, below, and behind the stage provided openings into non-diegetic spaces that also intersected with spaces of performance, as in the balcony scene in *Romeo and Juliet*. As both an actor and shareholder in the playing companies for which he wrote, Shakespeare understood and experimented with these visual possibilities throughout his career, including his history plays, which employ specific modes of visuality to define monarchy and to understand nature as a structure of representation.[56]

As scholars have noted, Shakespeare was intensely interested in both philosophical and representational problems of vision throughout his plays and poems.[57] In sonnet 24 ("Mine eye hath played the painter"), for example, he employs a complex metaphor in which the speaker keeps the "form" of his beloved in the "table" of his "heart," while his body functions as the "frame" that holds the form. The art of "perspective" allows the lover to see that "form" of his beloved in his own heart by looking through her eyes. Perspective thus links the visible to the invisible through languages of the body, raising questions about the power of words in the poem to elicit spectacle. Elsewhere, Shakespeare links visual perspective with the difficulty of resolving visual cues, as when Bertram in *All's Well that Ends Well* describes the "scornful perspective" of contempt that controlled his eye as he first viewed the daughter of La Few, making all of her features appear "most hideous" in being "extended or contracted" (5.3.44–52). Shakespeare likewise alludes to the art of perspective in *Twelfth Night* (c. 1600) when the separated twins Sebastian and Viola appear for the first time on stage near the end of the play. Witnessing both twins together, Duke Orsino observes that they seem to be "one face, one voice, one habit, and two persons" according to a "natural perspective, that is and is not" (5.1.206–7). Orsino refers here to the visual deception produced naturally by landscape in particular types of light, where images seem to multiply for some viewers at certain angles. Sebastian's own response to seeing his sister dressed as a man also raises theological questions that further the metaphor: "I never had a brother, / Nor can there be that deity in my nature / Of here and everywhere" (5.1.216–18). In witnessing his own double, Sebastian jokes that only a form of divinity (or its manipulated opposite) could account for why he is present in multiple beings. These examples show how Shakespeare approaches perspective in and through problems of theology, epistemology, and representation. Shakespeare's history plays explore such problems with special intensity, especially in relation to the "mirror of princes" genre that he adapted in writing them.

Looking Awry in *Richard II*

Richard II overlays tragic structures on historical narrative to stage the fall of a divinely appointed English king.[58] Many characters in the play, including Richard himself, define themselves in and through modes of vision. Embracing his own divinity as monarch, for example, Richard identifies himself as a sun king who both gazes out at his subjects and receives their gaze in return. He alludes to this visual divinity most fully when the Duke of Aumerle warns him about the threat that Bolingbroke poses to his reign. Richard responds:

> Discomfortable cousin, knowest thou not
> That when the searching eye of heaven is hid,
> Behind the globe that lights the lower world,
> Then thieves and robbers range abroad unseen
> In murders and in outrage bloody here.
> But when from under this terrestrial ball
> He fires the proud tops of the eastern pines
> And darts his light through ev'ry guilty hole,
> Then murders, treasons and detested sins,
> The cloak of night being plucked from off their backs,
> Stand bare and naked, trembling at themselves?
> So when this thief, this traitor Bolingbroke,
> Who all this while hath reveled in the night,
> Shall see us rising in our throne, the East,
> His treasons will sit blushing in his face,
> Not able to endure the sight of day,
> But, self-affrighted, tremble at his sin.
> Not all the water in the rough rude sea
> Can wash the balm off from an anointed king. (3.2.32–50)

Richard here imagines his power as monarch in terms of a divine and cleansing light that illuminates the natural world, alternating between images of sunlight and water to justify his apparent immunity from rebellion. The sun, even while hidden away at night, purges traitorous rebellion, according to Richard, by use of a divinely antiseptic light. In his own words, Richard defines himself as both a spectacle of divine power and an agent of surveillance. The play, in fact, mixes both older and newer models of kingship, including a form of panoptic power that Richard describes here.[59] Richard aligns Bolingbroke with the treasonous darkness but also suggests that his treasons will become visible in the blush of his face once this divine light shines on him at daybreak when the royal throne rises in the east. Bolingbroke and other rebels to the throne aim, for their part, to "redeem…the blemished crown" and to "make high majesty look like itself" (2.1.293, 295).

Intensely visual language also shapes the central confrontation between Bolingbroke and Richard at Flint castle, where Richard hopes to return to his former glory by becoming an object of the collective gaze. The king's faith in his own authority collapses as he finds that he can become only a "glist'ring Phaëthon" who is destined to fall (3.3.178). Shakespeare marks this fall visually by having Richard descend from the upper balcony above the stage to the "base court" where kings themselves "grow base" (3.3.180). Bolingbroke's subsequent rise to power also incorporates models of vision, as reported by the Duke of York to his wife after witnessing his coronation as King Henry IV: "So many greedy looks of young and old" gazed at the new king through windows with "desiring eyes" upon his "visage" (5.2.13–15).

Some scholars have understood the fall of Richard II and subsequent rule of Bolingbroke/Henry IV as a transition from divine-right monarchy to a Machiavellian model of might makes right.[60] According to this interpretation, Richard's belief in the divinity of his office of king divorces him from political reality and makes him vulnerable to the Machiavellian manipulations of Bolingbroke. Such analysis makes sense in terms of the political plots of *Richard II* and the Henry plays, but it overlooks the afterlife of divine monarchy in the figure of Henry VI and the hybrid forms of Machiavellian and divine politics throughout the history plays. The historical chronicles by Edward Hall and Raphael Holinshed that Shakespeare used to write his history plays are especially important in this regard. The second edition of Holinshed in particular includes accounts of the topography of the island of Britain together with its religious and natural history. One narrative that Holinshed foregrounds is that by which England bore witness to the establishment of the Christian church as early as 177 CE. After the new religion arrived in Britain at such a time, according to Holinshed, the gospel "was freely preached without inhibition of her prince."[61] His chronicle also features geographic descriptions of the islands surrounding Britain's coastline, lists of the names of the rivers and streams, accounts of the quality of the air and soil, and a section on "the general constitution of the bodies of the Britons." In the latter section, Holinshed praises British men as "white in color, & strong in body"—as warriors celebrated for the bravery in fighting and winning battles in France, Germany, and elsewhere—while labeling those of more southern climes crafty and "subtle."[62]

These passages indicate the power of divine providence to shape the nations of the world and its people. In his description of the deposition of Richard and the election of Henry IV to the status of monarch, for example, Holinshed notes in his second edition of the *Chronicles* (1587) that "in this dejecting of the one, and the advancing of the other, the providence of God is to be respected, and his secret will to be wondered at."[63] Shakespeare inscribes this providential language in the order of nature in *Richard II* when Bolingbroke describes his confrontation with Richard as a combat between the four elements: "Methinks King Richard and myself should meet / With no less terror than the elements / Of fire and water" (3.3.54–6). Such "elements" of nature echo the earlier and more famous passage in the play when Gaunt celebrates the divine nature of England itself:

> This royal throne of kings, this sceptered isle,
> This earth of majesty, this seat of Mars,
> This other Eden, demi-paradise,
> This fortress built by Nature for herself
> Against infection and the hand of war [...].
> This blessèd plot, this earth, this realm, this England,
> This nurse, this teeming womb of royal kings,
> Feared by their breed and famous by their birth,
> Renownèd for their deeds as far from home—
> For Christian service and true chivalry—
> As is the sepulcher in stubborn Jewry
> Of the world's ransom, blessèd Mary's son,
> This land of such dear souls, this dear, dear land. (2.1.40–44, 50–57)

Gaunt celebrates the island nation of England as a defensive "fortress built by Nature" that protects it from war and bloodshed, a blessed origin that gives England its reputation for Christian chivalry. Alluding to both classical and Christian deities, he portrays England as a space of godly blessing and global power. His speech shifts the location of authority away from the struggle for power between Richard and Bolingbroke and toward the land itself, which he defines in terms of the triumph of Christianity over "stubborn Jewry." Drawing on 1 Timothy 2:5–6 ("the man Christ Jesus, who gave himself a ransom for all men"), Gaunt imagines England as both a sacrificial lamb for the world and as a "seat of Mars," a force of military conquest against the "triumphant sea." This nationalist chivalry includes military crusades to recover the "sepulcher" of Christ from the Muslims. For Gaunt, it is the nation of England itself, divinely created by Nature, that should be protected and cherished as a source of legitimate power as opposed to any particular king who reigns at a given time.

The prophetic death of Gaunt shortly after this speech also introduces the arts of perspective into the play. Anticipating some "unborn sorrow, ripe in Fortune's womb" (2.2.10), the Queen betrays her anxiety to one of Richard's attendants, Bushy, who attempts to mollify and perhaps silence her by explaining that her griefs multiply.

> Like perspectives which, rightly gazed upon,
> Show nothing but confusion; eyed awry,
> Distinguish form. So your sweet Majesty,
> Looking awry upon your lord's departure,
> Find shapes of grief, more than himself, to wail,
> Which, looked on as it is, is nought but shadows
> Of what it is not. (2.2.18–24)

In the manner of Holbein's celebrated portrait of *The Ambassadors* (1533) and an anamorphic portrait of Edward VI by the painter William Scrots that Shakespeare may have known, the passage uses the language of visual perspective to describe political, personal, and emotional confusion.[64] The eyes see only confused visions when looking directly on a perspectival image. One must look awry in order to find "form" in the image, suggesting the need for multiple viewing angles. For Bushy, the Queen's fears come to nothing and, like Lear, he believes that nothing can come of nothing. But the Queen is not convinced. She recognizes, like the play more broadly, that invisibility can be the condition of certain forms of personal and political agency.[65]

Both Richard and the Queen struggle to discern the reality beneath the appearance of those around him. Moreover, Richard tries but fails to make his own interiority visible to himself, and in his attempt to handle Bolingbroke and the other rebels he fails to see what is right in front of his eyes, even as he claims to be able to see treasons in the darkness of night. Such blindness comes to a head in the deposition scene, when Northumberland asks Richard to read a list of "grievous crimes" committed against the state in order to be "worthily deposed" (4.1.216, 220). Richard replies that their act of deposing him will itself make a "blot" in the "book of heaven" (lns. 229). Asked to read the articles of grievance against him, Richard refuses, noting that

> Mine eyes are full of tears; I cannot see.
> And yet salt water blinds them not so much
> But they can see a sort of traitors here.
> Nay, if I turn mine eyes upon myself,
> I find myself a traitor with the rest;

> For I have given here my soul's consent
> T'undeck the pompous body of a king. (4.1.237–43)

Where the Queen's tears fragment her vision like light passing through a kaleidoscope, Richard achieves a new perspective in his state of duress. Like the fall of Oedipus in Greek tragedy, his fall generates self-knowledge and spiritual insight. Following the action with their own eyes, Shakespeare's audience achieves a new and mixed perspective on the action before it. Monarchy is divided, split literally into two, and kingly power becomes more firmly enmeshed in the politics of visibility. Perspective is simultaneously personal, philosophical, political, and emotional.

In the midst of his own formal deposition, Richard asks for a "mirror" that will "show" him "what a face I have" (4.1.258–9). Importantly, this mirror transforms his image into a "book" in which he can read his own "sins" (lns. 267–8). Gazing at his own reflection, he shows surprise that he does not resemble one who is about to relinquish kingly powers: "Hath sorrow struck / So many blows upon this face of mine / and made no deeper wounds?" (lns. 270–72). Finding only "brittle glory" in his own reflection (ln. 280), the King dashes the mirror to the ground, breaking it into "an hundred shivers" that represent his own shattered self and the instability of the crown he is about to relinquish (ln. 282). As he discovers, the mirror proves inadequate as a tool for reading the book of the self.

The defeated Richard also casts himself in the role of a crucified Christ, accusing all who stand and "look" upon him of being like Pilate who show only "outward pity" even as they deliver him to the "sour cross" (4.1.230–34). The Abbot of Westminster describes the scene in an older theatrical lexicon as a "woeful pageant" (ln. 314). Richard's exit from the role of king accompanies his desire to be invisible, away from the "sights" of all (ln. 308). The play grants his wish by enacting his murder off stage. Bolingbroke, who in Shakespeare's version of the story does not order the death of Richard, ends the play wracked by guilt with his soul "full of woe" (5.6.45). He vows to make a pilgrimage to the Holy Land to "wash this blood off from my guilty hand" (5.6.50), yet as Henry IV he never escapes the guilt of the original sin of deposition, which also hangs over his descendants on the English throne. Bolingbroke's strong visual presence enables his rise to power at the expense of Richard's fractured self but his assumption of the throne constructs a new visual model of monarchy that tries and fails to suppress the hidden sin of deposition.

It is understandable that forms of visuality continue to shape the characters and politics of Shakespeare's tetralogy. Both Henry IV and Prince Hal invoke the sun in the two parts of *Henry IV*, for example, but they do so in different ways than Richard II. As a profligate prince who socializes with Falstaff, Bardolph, and others at the Boar's Head, Hal shuns the spotlight in pursuit of a more shadowy existence of private pleasure. Yet such self-concealing, we learn, is central to his plot to achieve power. At the end of his first scene in *1 Henry IV*, for example, Hal announces his plan to "imitate the sun" by allowing "base, contagious clouds" to cover him so that his "reformation, glittering o'er [his] fault" will show him more "goodly" and "attract more eyes" than if he had no "foil" for his spectacle of power (1.2.170–92). Hal's father, King Henry IV, reveals a similar strategy later in the play. Scorning his son for being an "alien" to the court and to his own blood line, the King narrates his own rise to power in a story that tellingly avoids discussing the deposition of his predecessor (3.2.34). Henry IV celebrates his capacity to "pluck allegiance from men's hearts" by virtue of his ability to control his own representation before his subjects (ln. 52). The elder Henry compares himself to a comet rarely seen and therefore more "wondered at" (ln. 47), enhancing his own "sun-like majesty" (ln. 79). Where Richard cannot control the conditions of his own royal spectacle, Henry IV and his son employ Machiavellian strategies in efforts to manage their visibility. But Henry IV experiences similar vulnerabilities as king as does Richard II. Haunted by his own role in the death and deposition of his predecessor, his power proves tenuous.

Perspectival Power in *Henry V*.

Chronicle histories on which Shakespeare drew in creating the heroic figure of Henry V celebrate the monarch as a "mirror" for all to gaze on. Hall, for example, describes the prudent counselors who helped to make King Henry "a singular mirror and manifest example of moral virtues and good qualities to his common people and loving subjects."[66] Such a monarch takes up a mimetic role toward the divine, as Erasmus notes in his *Education of a Christian Prince* (1516): "God set up a beautiful likeness of himself in the heavens, the sun," just as he "established among men a tangible and living image of himself, the King."[67] But the glory of monarchy for Hall is not solely dependent on the person of the king himself. Rather, it is the product of "rich regions, subjects, and beautiful cities and towns" which have been given to England and Ireland by God.[68] Henry's status as a true mirror to his people emerges in and through his military victories in France and his acceptance of the divine providence that makes such victories possible.[69] As the mirror of kings, Shakespeare's most powerful monarch grounds his power in modes of visuality that also depend on the divine order of nature.

But Henry can be an idealized mirror of monarchs only in so far as Shakespeare's theater can generate adequate images of him. Problems of visuality in the play suggest the challenges of such a project. As monarch, Henry inspires wide-ranging loyalty and duty from his followers, even though his actions and the bloodiness of war undercut the play's efforts to justify the invasion of France. As Norman Rabkin observes, "Shakespeare created a work whose ultimate power is precisely the fact that it points in two opposite directions, virtually daring us to choose one of the two opposed interpretations it requires of us."[70] But this duality of Henry as a figure of both Machiavellian calculation and authentic heroism generates a bifurcated, shifting, and multi-faceted character who refuses unifying interpretations.[71] Like the anaphoric skull at the bottom of *The Ambassadors*, the play challenges its audience to sustain both interpretations of Henry at the same time, even as its mode of representation continuously seems to defeat such efforts.[72]

Hal's journey across the two parts of *Henry IV* toward the throne, moving from profligate tavern-dweller to divine monarch, depends to a significant extent on his capacity to control his own legibility as a character. Before he becomes king, he engages in a carefully orchestrated plan to reveal himself only partially. Once king, he (like all monarchs before and after) becomes subject to greater public scrutiny. Shakespeare's own staging of this ascent also engages in acts of theatrical hide and seek. The Chorus appears first in *Henry V*, apologizing for the limitations of the capacity of the stage to represent the "brightest heaven of invention" (1.0.2) and imploring the audience to employ its "imaginary forces" to comprehend what the theater cannot stage (ln. 18). The next scene introduces the monarch at the center of the play indirectly through the dialogue of the Bishops of Canterbury and Ely. Ely describes Henry's glittering reformation as if an angel arrived to whip the "Adam" out of him, making him "all-admiring" and capable of "reason in divinity" (1.1.38–9). For his part, the Bishop of Ely employs a natural analogy to describe Henry's alteration:

> The strawberry grows underneath the nettle,
> And wholesome berries thrive and ripen best
> Neighbored by fruit of baser quality;
> And so the Prince obscured his contemplation
> Under the veil of wildness, which, no doubt,
> Grew like the summer grass, fastest by night,
> Unseen, yet crescive in his faculty. (1.1.60–66)

In the same way that the strawberry conceals itself under baser weeds, Ely suggests in his horticultural analogy, so too did Henry surrounded himself with what Canterbury has described as the "unlettered, rude, and shallow" inhabitants of Eastcheap (ln. 55). Henry's ascent to the throne allowed his own invisible powers of "contemplation" to grow themselves like summer grass "unseen" until it became part of his "faculty" as a ruler. Shakespeare's theater has given us glimpses of this faculty in the two parts of *Henry IV*, but the full measure of such greatness has not been recognized until the unveiling of the prince's fullest political powers as king in *Henry V*.

Obscured by the lengthy monologue of Canterbury justifying war with France on the basis of Salic law, Henry emerges sun-like at the end of 1.1 in his rousing call to arms before the French ambassador and the gift of tennis balls from the Dauphin.[73] In claiming that he will "be like a king" and "show" his "sail of greatness," Henry vows to "dazzle all the eyes of France" and strike the offending Dauphin "blind" with the very balls that he has sent as an insulting joke about Henry's wild youth (1.2.275–81). Unlike Richard, however, Henry does not confine his power to his own capacity to be seen. Rather, he develops a broader strategy of warfare that incorporates his own soldiers and the order of nature. In the Battle of Harfleur, for example, he urges his men to become better soldiers by means of mimesis that centers on vision:

> Once more unto the breach, dear friends, once more,
> Or close the wall up with our English dead.
> In peace there's nothing so becomes a man
> As modest stillness and humility;
> But when the blast of war blows in our ears,
> Then imitate the action of the tiger:
> Stiffen the sinews, conjure up the blood,
> Disguise fair nature with hard-favored rage.
> Then lend the eye a terrible aspect
> …

Be copy now to men of grosser blood
And teach them how to war. And you, good yeomen,
Whose limbs were made in England, show us here
The mettle of your pasture; let us swear
That you are worth your breeding—which I doubt not,
For there is none of you so mean and base,
That hath not noble lustre in your eyes.
I see you stand like greyhounds in the slips,
Straining upon the start. The game's afoot:
Follow your spirit, and upon this charge
Cry, 'God for Harry! England, and Saint George!' (3.1.1–9, 24–34)

Shakespeare here invokes modes of mimesis within the order of nature in order to stage the political drama of war. The ultimate sign of military prowess for Henry is the ability of soldiers to become like animals—to "lend the eye a terrible aspect" and to gain "noble lustre" in the eye in order to "disguise fair nature with hard-favored rage." Thus successful soldiers should 'imitate the action of a tiger' by stiffening their arms and making physical changes to their body." Like John of Gaunt and Richard II, Henry defines his power in terms of "nature," including a "gallèd rock" (lns. 8, 12) and the "wild and wasteful" ocean (lns. 14), in addition to the racing greyhounds in this speech. Henry's followers must comport themselves into new and unnatural forms to "show" themselves as a "brass cannon" and other instruments of war. Ever conscious of social status, Henry also urges them to imitate their betters, since better-born soldiers are supposed to be a "copy" to those of "grosser blood."

The transition from *Richard II* to the Henry plays also accompanies a heightened degree of artificial representation. The divinized natural world that Gaunt and Richard II both invoke to justify English monarchy remains under the surface, but it is wrapped in a new language of visual artifice and heightened attention to theatrical representation. The Chorus who appears in the folio version of the play foregrounds visuality and outlines a model of audience participation that situates visual representations of the theater in relation to the other senses.[74] According to this Chorus, the act of watching the performance demands active participation on the part of an audience that must supply what the stage cannot by employing the force of imagination. Because the "flat unraised spirits" treading the "unworthy scaffold" of the stage do not have the power to represent

Harry as he should be seen in the form a god of war, the Chorus invites and instructs the audience to join in speculative and collaborative vision as they watch, with a possible allusion to classical accounts of vision in Empedocles and in Plato's *Timaeus* that regard the eye as made of fire (45b). Such participatory and voluntaristic vision is entirely expected for early modern audiences and perhaps quite foreign to modern ways of viewing theater. The Chorus demands that the audience use its speculative powers of imagination, reminding us of our need to "work" our "thoughts" so as to "see" events that cannot be staged, such as the siege of Harfleur (3.0.25). The Chorus also implores the audience to "piece out" the "imperfections" of the theatrical representation so as to construct the history in collaboration with the actors (1.0.23). In the process, the play draws attention to problems of representation that beset Shakespearean theater more broadly, reminding the audience not just of the unrepresentability of war but also the unrepresentability of Henry's own divine kingship.

The Chorus defines visibility not simply in relation to the operation of light on the eye, but rather in a fuller sense as the theatrical production of language, ideas, objects, and images reinforced by the power of the imagination. In this sense, the Chorus encourages the audience to see what is not there, and even to see what is not possible to be there. For example, before act 3, he urges audience to "behold the threaden sails" of the English ships as they cross the channel toward France, "borne with th'invisible and creeping wind" (3.0.10–11). Such "threaden sails" might be represented visibly on stage, but the "creeping wind" could never be seen directly, even if it could be harnessed as part of the stage spectacle. Like the God in apophatic theology who is not capable of human representation, the unrepresentable divinity of Henry as monarch can only be known in and through the language, bodies, and images that are materially present on the stage. The comic figure of the Welsh Captain Fluellen comically embodies this limitation. Appearing out of nowhere in act 3 to urge Pistol, Nym, and others unto the "breach" to fight (3.2.19), he is mostly concerned throughout the drama with the relation between the actual war and the theoretical Roman "disciplines" that proscribe the ideals of warfare. When he encounters the Irish Captain Macmorris and the Scottish Captain Jamy outside of the walls of Harfleur, Fluellen interrupts the action to beseech Macmorris to engage in some "disputations" about the Roman wars (3.3.37). In this gesture, which Macmorris churlishly dismisses, Fluellen draws attention to discrepancies between the abstract ideals of warfare and the messy reality of actual combat. He is also a figure around whom action and ideas not visible to the audience can coalesce. His involvement with the practical joke of the leek that Pistol must display on his hat and the glove that King Henry gives him to wear as a joke to incense Williams continues this work as a mode of comic relief to offset the bloody violence of combat that consumes the final acts of the play.

The structure and importance of visibility take a new turn in the play following the defeat of the French at Agincourt, when Henry V attempts to woo the French princess Katharine to be his bride. In a comic version of the incapacity of Shakespeare's stage to represent Henry as Mars, the French Duke of Burgundy jokes that if Henry could "conjure" the "true likeness" of the Cupid in Kate, she would blush with virgin modesty because Cupid is normally naked (5.2.270). Henry replies flippantly that all maids ultimately "blink and yield, as love is blind and enforces" (ln. 275). As he turns to members of the French court, he boasts that they should "thank love for my blindness," since such blindness prevents him from spying "many a fair French city for one fair French maid that stands in my way" (ln. 289–91). The French King joins the uncomfortable joke by noting that Henry sees his daughter and the French cities "perspectively," meaning that "the cities turned into a maid—for they are all girdled with maiden walls that war hath never entered" (lns. 292–94). Henry responds to his future father-in-law's observation that Henry has seen both Kate and the as-yet-unconquered French cities "perspectively" by noting that the "maiden cities you talk of may wait on [Kate]: so the maid that stood in the way for my wish shall show me the way to my will" (lns. 297–99).

It is important to note that Shakespeare gives Kate no voice in this exchange. Her body becomes a vehicle of Henry's desire to sack "many a fair French city" and thus to fulfill his "will." This sexualized and violent joke between powerful men at court reveals that Henry's act of conquering France, which he earlier justified in legal and Christian language, may ultimately be nothing more than a masculinized will to power. The all-male laughter of these punning male ministers also reveals and conceals the violent conquest that has preceded the final peace in the play, suggesting fissures in the wished-for unity between both the kingdoms and the newly married couple. Where Henry jokes that the virgin Kate has obscured his power to see other French towns he might invade, the King conflates his daughter with those towns on the basis of the sexist logic that both are virgin territory. The language of perspective here links the "maiden walls" of Kate to the unconquered French towns (5.2.294), suggesting historical links between perspective glasses and the land that they surveyed while also invoking fantasies of colonial possession of feminized land. Such a metaphor erases Kate's identity as a human being and, like Julia and Silvia in the closing scene of *Two Gentlemen of Verona*, she remains silent for the remainder of the play.

This crass joke on perspective in the final lines of *Henry V* also provides a fitting conclusion to a play that consistently foregrounds the problems of vision. The model of monarchy to which Henry V returns is that of the "sceptered isle" of John of Gaunt's speech in *Richard II* in which the land itself is a source of divine power. The union of Kate and Henry, overseen by what the Queen of France calls "God, the best maker of all marriages" (5.2.328), produces a "Christian-like accord" (ln. 322) between man and wife and between England and France to ensure a royal treaty or "paction" (ln. 334) that will hopefully produce long-lasting peace between the nations. Such hope captures Gaunt's definition of England as a "teeming womb of royal kings." But the brief Epilogue spoken by the Chorus at the very end of the play recalls what Shakespeare's stage has "shown" to audiences of his history plays—that the next member of the royal blood, Henry VI, will go on to lose the territories in France that his father won (Epilogue, ln. 13). This jarring admission is a final whiplash of perspective for spectators who have followed the ups and downs of Henry's own reign.

These examples illustrate some of the ways that Shakespeare links questions of vision to problems of knowledge, nature, and divinity in his portrait of England's most celebrated monarch. Shakespeare's multi-faceted portrait of Henry V produces a multi-perspective theater that mixes older and newer models of vision, organizing the senses by outlining new relations between vision, knowledge, and being. In the process, Shakespeare invites a mode of diverse witnessing of English history by the audience, producing a form of what Lorraine Daston and Peter Galison call "collective empiricism." For Daston and Galison, collective empiricism as practiced by the Royal Society in the seventeenth century encouraged skeptical responses to accepted truths. Their examples of the "Epistemologies of the Eye" produced by such vision include shared "repositories of images of record for the observational sciences" by natural philosophers around the globe.[75] These repositories of images generate maps as modes of representation that are collectively created in order to stabilize unstable systems of knowledge.

Shakespeare's history plays generate their own repositories of images that challenge any unified account of English history. At the same time, they stage acts of seeing as simultaneously physical, material, and political.[76] But Shakespeare's collective empiricism is de-centralizing, especially in *Henry V*. Due in part to the repeated conjuring work of the Chorus, the play invites reflection on its own status as a work of theatrical spectacle. Such self-reflection shapes how Shakespeare both embraces and critiques ideas of divine monarchy and divine creation of the natural world. In a larger sense, the playhouse participates in a move away from the centrality of emblematic and iconic images that are seen to replicate or partake of divinity or the real and toward a more "modern" sense of the construction of images that are self-aware of their status as images. This self-awareness also becomes the imperative of spectators trained to pay particularly close attention to visual details of performance, a capacity likewise embraced by early modern theologians and natural philosophers who sought to understand the order of nature.

Notes

1. Stuart Clark, *Vanities of the Eye*, 9–38. Chapter 7 also analyzes untrustworthy vision in *Macbeth*.
2. See, e.g., Armelle Sabatier, *Shakespeare and Visual Culture: A Dictionary*.
3. Challenging the claim by John Orrell that the Globe playhouse was "an acoustical auditorium, intended to serve the word and the ear more fully than the image and the eye," together with historical assumptions by Andrew Gurr and others that playwrights preferred playgoers who were auditors rather than spectators, Gabriel Egan in his digital survey of a broad database of works by literary, dramatic, and prose authors concludes that "plays were much more commonly thought of as visual rather than aural experiences in the literary and dramatic writing of the period." See "Hearing or Seeing a Play, 332." See also Orrell, *The Quest for Shakespeare's Globe*, 140.
4. Timothy Reiss, *The Discourse of Modernism*, 30. Emphasis in original.
5. Ibid., 25.
6. Such a claim differs from the approach taken in the essays collected in *Visual Cultures of Science*, ed. Pauwels.
7. Peter of Limoges, *The Moral Treatise on the Eye*, 36.
8. Reisch, *Margarita philosophica,* 176, 174.
9. *Species* were also etymologically linked to "spectacle," including perspective, speculation, and introspection.

10. A. Mark Smith, *From Sight to Light.*
11. See Johannes Hoff, *The Analogical Turn*, 36–7.
12. For an account of the importance of the crystalline humor to Vesalius's model of vision, see Vincent Ilardi, *Renaissance Vision from Spectacles to Telescopes*, 242.
13. See Felix Platter, *De corporis humani structura et usu*, 186. See also Katrien Vanagt, "Early Modern Medical Thinking on Vision and the Camera Obscura," 571.
14. See, e.g., Meredith Ziebart, "Cusanus and Nominalism," 227.
15. Pierre de La Primaudaye, *The Second Part of the French Academy*, 4. On Shakespeare's affinity for La Primaudaye and his use of the text in writing *Love's Labor's Lost*, see Rolf Soellner, *Shakespeare's Patterns of Self-knowledge*, 79 and Dennis Taylor, *Shakespeare and the Elizabethan Reformation*, 146. Neither study mentions La Primaudaye's status as a prominent natural philosopher.
16. La Primaudaye, 72.
17. Ibid., 75.
18. For an overview of lens making in the seventeenth-century, which includes a section on Huygens, see D. Graham Burnett, "Descartes and the Hyperbolic Quest."
19. A copy of Galileo's letter is held in Special Collections at the University of Michigan. A digital image of the letter with translation and commentary is available at https://www.cabinet.ox.ac.uk/uses-telescope-1609-10, from which this translation has been taken.
20. See, e.g., W. Patrick McCray, *Glassmaking in Renaissance Venice.*
21. Leonard Digges, "Preface to the Reader," in *A Geometrical Practice, Named Pantometria*, sig. A3v.
22. See Paul Kocher, *Science and Religion in Elizabethan England*, 164–5 and Robert Goulding, "Thomas Harriot's Optics."
23. Bacon, *Novum Organum*, ed. Jardine and Silverthorne, 170. See also Harrison, "The Book of Nature," 4.
24. Robert Boyle, *Some Considerations Touching the Usefulness of Experimental Natural Philosophy*, 4.
25. John Dubreuil, *Perspective Practical.* The text was reprinted by a different publisher in 1698.
26. Bacon, *Novum Organum*, ed. Jardine and Silverthorne, 42.
27. Robert Hooke, *Micrographia*, 193.
28. Ibid., sig. a.
29. Ibid., 91.
30. Ibid., 155.
31. See Harrison, "The Book of Nature and Early Modern Science," esp. 25–6.
32. Reginald Scot, *The Discovery of Witchcraft*, 316.

33. Margaret Cavendish, *The Blazing World and Other Writings*, 142–44. For a feminist reevaluation of Cavendish's skepticism toward Baconian science, see Eve Keller, "Producing Petty Gods: Margaret Cavendish's Critique of Experimental Science."
34. As Brandie R. Siegfried and Lisa T. Sarasohn note in their introduction to an important anthology of essays on Cavendish, "the natural world, human nature, and the Divine were all intimately connected for Cavendish and her contemporaries," but they disagreed about how and whether scientific experiments could access the divine origins of the *ordo creationis*. See their "Introduction" to *God and Nature in the Thought of Margaret Cavendish*, esp. 7–8.
35. As a set of techniques for representing three-dimensional space in two dimensions, the arts of perspective are quite ancient. Art historians have noted the limited use of perspective, for example, in certain Roman murals and as motifs in medieval artists. It was only in the middle of the fifteenth century in Florence, however, that artists developed a more unified approach to perspective that sought to create a single, overarching vanishing point.
36. For an account of the metaphysics of Alberti's theory of linear perspective, see James S. Ackerman, "Alberti's Light."
37. For a literary example of the mirroring power of language, see Andrea Schutz, "Absent and Present Images."
38. Ackerman, for example, understands Albertian perspective in contrast to Neoplatonism in *Origins, Imitation, Conventions*, 10–11. Likewise, Karsten Harries reads Alberti's theory of perspective in terms of the Cartesian revolution of science that celebrates "reason and nature" over divinity in *Infinity and Perspective*, 81.
39. Leon Battista Alberti, *On Painting*, 47.
40. Charles H. Carman, *Leon Battista Alberti and Nicolas Cusanus*, 162.
41. Euclid, *The Elements of Geometry*, sig. b1.
42. Alberti, *On Painting*, 34.
43. Sidney, *The Defense of Poesy* (c. 1579), 236.
44. George Puttenham, *The Art of English Poesy*, ed. Whigham and Rebhorn, 110.
45. Rayna Kalas, *Frame, Glass, Verse*, 137.
46. Diehl, p. 73.
47. Calvin, *Commentary on the Acts of the Apostles*, trans. John W. Fraser, in *Calvin's Commentaries*, 7:104. Cf. Diehl, *Staging Reform*, 70, who nevertheless misquotes Calvin slightly.
48. Waldron, chpt. 1. See also Diehl and Knapp.
49. Somewhat surprisingly, there are relatively few studies of the early modern stage in relation to optics. Philip Armstrong offers a postmodern and psy-

choanalytic account in *Shakespeare's Visual Regime: Tragedy, Psychoanalysis and the Gaze*. Mary Thomas Crane suggests in a more historically grounded and astute chapter on "Optics" that "Early modern playwrights seem to have been fascinated by contemporary developments in optics but wary of the theatrical potential of optical effects." See "Optics" in *Early Modern Theatricality*, ed. Turner, 268. She provides a brief survey of classical and medieval theories of vision, with readings of Robert Greene's *Friar Bacon and Friar Bungay*, Jonson's *Alchemist*, Shakespeare's *Tempest*, and Thomas Tomkis's *Albumazar*.

50. Archaeological discovery of the foundations of the Globe and the Rose theaters in the 1980s did prompt some new studies. See, e.g., Julian Bowsher, *Rose Theatre: An Archaeological Discovery* and his *Shakespeare's London Theatreland*.
51. See John Astington, *Stage and Picture in the English Renaissance*, 197.
52. See Peter Thompson, *Shakespeare's Theater*, 2d ed., (New York: Routledge, 1992), 41–2.
53. First envisioned by the American actor Sam Wanamaker in the 1970s and designed by the architect Theo Crosby, the reconstructed Globe theater depended on the labor and knowledge of architects, archaeologists, theater historians, carpenters, and members of historic building societies.
54. These estimates come from John Orrell, "Designing the Globe: Reading the Documents," in Mulryne and Shewring, eds., 54–55.
55. Andrew Gurr, "Shakespeare's Globe: A History of Reconstruction," in Mulryne and Shewring, eds., 43–44.
56. On Shakespeare's history plays as a challenge the authority of monarchy by submitting it to new perspectives of early modern "subjects," see David Scott Kastan, "Proud Majesty Made a Subject," 459–475.
57. There are many examples of studies of Shakespeare's intense interest in visuality. For studies of his interest in optics in particular and Italian developments in visual technologies, see the essays in *Shakespeare and the Visual Arts*.
58. See., e.g., Christopher Pye, *The Royal Phantasm* which is based on his earlier essay "The Betrayal of the Gaze." See also Richard Ashby, 'Pierced to the Soul'.
59. Ashby, "Pierced to the Soul," 205.
60. A classic version of this reading can be found in Irving Ribner, "Bolingbroke, a True Machiavellian." For a more modern and nuanced approach, see Hugh Grady's section on "The Discourse of Princes in *Richard II*," in *Shakespeare, Machiavelli, and Montaigne*, 58–81.
61. Raphael Holinshed, *The First [Last] Volume of the Chronicles of England, Scotland, and Ireland*, fol. 9v.
62. Ibid., fol. 39.

63. Raphael Holinshed, *Holinshed's Chronicles*, ed. Hansen and Wallace, 24.
64. The image of Edward VI was in Whitehall in the late sixteenth century, as at least one traveler recalls. See Ernest B. Gilman, *The Curious Perspective*, 55.
65. For an elaboration of this point with reference to this passage, see Scott McMillin, "Shakespeare's Richard II: Eyes of Sorrow, Eyes of Desire," 41–3.
66. Edward Hall, *The Union of the Two Noble and Illustrious Families of Lancaster and York*, sig. f1v.
67. Desiderius Erasmus, *The Education of a Christian Prince*, in *Erasmus: The Education of a Christian Prince*, ed. Jardine, 23–24.
68. Hall, *Union*, sigs. f3v-f4.
69. Ibid., sig. f2v.
70. Norman Rabkin, *Shakespeare and the Problem of Meaning*, 34.
71. The contrast between a secular, Machiavellian strain of politics in the play and a more conventionally providential understanding informs many accounts of *Henry V*, including Stephen Greenblatt's "Invisible Bullets" in *Political Shakespeare*, 18–47, and Phyllis Rackin, *The Stages of History*. For more on the influence of Machiavelli in early modern England, see Felix Rabb, *The English Face of Machiavelli*.
72. Gilman makes a related claim about the duality of monarchy in *Richard II* in his chapter on "*Richard II* and the Perspectives of History" in *The Curious Perspective*, 88–128. See also Pye, *Regal Phantasm*, 14.
73. This is not to deny the significance of the defense of war, especially in relation to gender in the play. See Katherine Eggert, "Nostalgia and the Not Yet Late Queen."
74. The Chorus also seems to apologize for divine monarchy. As Michael Goldman notes, "the figure of the Chorus rousing the audience to cooperation and excitement is rather like the figure of Henry addressing his men." See *Shakespeare and the Energies of Drama*, 61.
75. Daston and Peter Galison, *Objectivity*, 22. See also Daston, "Baconian Facts, Academic Civility, and the Prehistory of Objectivity."
76. Here I agree with Evelyn Tribble, who argues that "sight is simultaneously a physical, biological, cognitive, material and cultural phenomenon" in her chapter on "Sight and Spectacle" in Karim-Cooper and Stern, eds. 237.

CHAPTER 6

Natural Medicine in *All's Well that Ends Well* and *Macbeth*

Like optics, medicine (or "physic") participated in the return to nature in early modern Europe. Yet early modern physic raised additional questions and problems. Medicine was subject to state control, for example, and involved a wider range of beliefs, practices, and modes of representation than optics. As Nancy Siraisi notes, early modern medicine entailed "a diverse collection of areas of knowledge and of intellectual interests and commitments" that were "pursued in the context of a variety of institutions, occupations, skills, and activities."[1] Like optics, however, early modern medicine was deeply enmeshed in a variety of religious contexts and theories. Even as universities began to offer medical instruction beginning in the thirteenth century, for example, European monasteries and abbeys remained primary sources of medical knowledge and treatment. And what we now call psychological care was also performed primarily by priests and clerical healers who focused on spiritual as well as emotional problems. In elite circles, medicine was increasingly professionalized in the sixteenth century, even though most acts of healing still took place in domestic settings, largely undertaken by women using herbal remedies passed down both orally and in written form over generations.[2] The explosion of printed books and treatises on medicine beginning in sixteenth-century England also helped to disseminate both old and new medical recipes and practices, giving rise to new debates about medicine and its relation to a divine order of nature.

A. Kitch, *Shakespeare's Theater of Nature*, Palgrave Studies in Literature, Science and Medicine,
https://doi.org/10.1007/978-3-031-78082-0_6

Both trained and untrained healers alike engaged in close observation of the natural world. In his *Brief Declaration of the Worthy Art of Medicine* (1586), for example, the English physician Thomas Gale urged doctors to study the "herbs, trees, stones, metals, minerals, beasts, fouls, fishes, and all other things, that creepeth on the face of the earth," created by "the high and mighty Lord."[3] As the historian John Harley Warner notes, early modern medicine was essentially a matter of "efforts to organize knowledge about nature."[4] Yet good doctors could not rely on empirical observation alone—they also had to consider invisible forces such as the soul. Moreover, physical symptoms of illness were notoriously deceptive. Successful healers had to understand subtle differences between different signs of illness and interpret those signs in relation to larger patterns. For this reason, Plato in the *Phaedrus* (c. 370 BCE) had warned against "hacks" and "empirics" who failed to link individual symptoms to larger patterns of disease (268a-269c). Likewise, Galen cautioned doctors who were led by their eyes alone.

Galen produced an enormous corpus of medical texts that remained foundational for medical practice and education well into the seventeenth century. Many of his writings were widely translated, debated, and used as teaching manuals. But an influential critic of Galen and the Aristotelian philosophy he largely followed emerged in the sixteenth century in the person of the alchemist, theologian, and healer Paracelsus. Where Galen defined the healthy body in terms of the abstract balance of its four humors—blood, phlegm, black bile, and yellow bile—Paracelsus believed that illnesses targeted specific parts of the body. Despite these differences, however, both Galen and Paracelsus grounded their medical writings and practices in theology.[5] Galen for instance believed that the art of medicine depended on an "omnipotent God" who invited direct observation of nature, while Paracelsus insisted that the bible was the root of "all philosophy and natural science."[6] Paracelsus understood medicine as a covenant between the healer and God, while Galen praised medicine itself as divine. Not everyone was convinced of these claims, however. Both the Presbyterian Thomas Hall and the Nonconformist Richard Baxter accused Paracelsus of trafficking with devils in the seventeenth century, for example.[7]

The contrasting medical ideas of Galen and Paracelsus shaped the art of physic throughout early modern Europe, even though healing practices were unevenly and imprecisely implemented and regulated.[8] In England, the Royal College of Physicians, the Barber-Surgeons Company, the Bishops of London, and aldermen of the City of London all monitored medical treatment, although they did so inconsistently. Most university-trained members of the Royal College followed Galen in proscribing "internal" remedies for the body that utilized some herbal remedies, while the Barber-Surgeons Company, which was formalized in 1540, performed procedures that required direct access to the body, including bloodletting and amputation. The Royal College policed medical practitioners, especially the itinerant and often female "empirics" who produced homemade herbal remedies.[9] One of its earliest statutes required the "declaration of secrets or 'arcana'" related to medical treatment. This effort to render public the secrets of household remedies passed down by oral traditions and Paracelsian treatments on the margins of established medicine participated, as we have seen, in a larger effort in the period to understand the order of nature through forms of representation.[10] In theory, the more elite doctors of the Royal College were distinguished from the manual practitioners of the Barber-Surgeons company and others who worked directly with the body to produce material remedies for illness and disease. But in practice, as Deborah Harkness has shown, early modern healers often employed both Galenic and Paracelsian remedies simultaneously, while many members of the Royal College depended on the manual labor of such "empirics."[11]

This chapter resituates early modern debates about medicine, including the conflict between Galen and Parcelsus, in terms of renewed efforts to understand the order of nature. It also demonstrates how Shakespeare's representations of the art and practice of medicine in *All's Well that Ends Well* and *Macbeth* build on classical, medieval, and early modern debates in medical writings. Helen in *All's Well that Ends Well* is a divinely inspired Paracelsian healer whose acts of healing bridge visible and invisible worlds. The "notorious" fistula (1.1.32) of the French King that she cures also becomes as a metonymy for Helen's own virginity—as her dialogue with Paroles early in the play suggests—and for the mysterious treatment of the disease itself, which occurs off stage. Likewise, the medical treatment of Lady Macbeth by a Doctor and Gentlewoman raises questions about the divine nature of disease and its legibility in *Macbeth*. I frame my reading of the tragedy through the eyes of the London astrologer and healer Simon Forman's famous notes on the play, which intersect with his own medical practice in ways that scholars have not appreciated.

The Natural Body

Pathological diagnoses for Galen demanded the interpretation of four overlapping sign systems: the nosological definition of a malady, the etiology of that illness, prognosis, and therapeutics. Medical observation included training in how to read and understand the body in order to achieve all four of these practices and to move from discerning symptoms in the body to diagnosing particular treatments for illness.[12] Galen argued that the central goal of healing was to remove all things contrary to nature (*contraria contrariis*) in order to preserve that which is natural.[13] He suggested treating heat with cold, for example, and curing "superfluous" humors with their opposite in an effort to achieve "dimunition."[14] By contrast, Paracelsus devised his own multi-tiered system of corporeal interpretation: chiromancy (palm and foot reading), physiognomics (face reading), "substantia" (reading the body as a whole), and interpreting manners and gestures (120). As these examples suggest, his approach to treating illness focused on the hermeneutics of the body as part of a new interest in nature as a system of representation. While Paracelsus agreed with Galen that direct observation was necessary but not sufficient for cure, moreover, he differed with Galen in proscribing treatments based on the logic of likeness or similarity between remedy and illness (*similia similibus*). Paracelsus also went further than Galen in understanding the role of what he called divine "secrets" (*arcanum*).[15] Those who had the gift of inner sight could detect the underlying structures of divinity beneath the signs and signatures of God's creation.[16]

Before studying the "book" of human beings, the physician for Paracelsus first had to master two other books—"*Sapientia*," the book of God, and the "firmament," or Book of Nature (84–5):

> We [...] discover everything that lies hidden in the mountains by external signs and correspondences, and thus also do we find all the properties of herbs and everything that is in the stones. There is nothing in the depths of the seas, nothing on the heights of the firmament, that man is unable to discover. No mountain, no cliff, is so vast as to hide or conceal what is in it from the eyes of man [*sic*]; it is revealed to him by corresponding signs...For each fruit is a sign, and through it we discover what is contained in that from which it stems. Similarly there is nothing in man that is not marked in his exterior, so that by the exterior one may discover what is in the individual

> who bears the sign [...] there are four ways by which the nature of man and of all living things can be discovered. [...] There is nothing that nature has not signed in such a way that man may discover its essence. (120–21)

In this passage, Paracelsus outlines the method by which God created all beings and material things out of a primordial "matrix." Like Aristotle, he compares God to a "woodcarver" (14) or a molder of clay for whom the earth provided the substance of such creation, to which He added form and "character" (18). Paracelsus also defines the human body as both material and "ethereal"—simultaneously earthly and divine—even as he acknowledges that these spiritual and material components of being inevitably conflict with one another: "the visible, material body wants one thing, and the invisible, ethereal body wants another" (41–2). Confronted with this internal spilt, there is no possibility for middle ground. Each part of the human attempts to exclude the other, leading to a constant state of internal war.

Paracelsus expanded on Galen's focus on natural medicine but promoted herbal and iatrochemical remedies for particular diseases rather than bloodletting and other strategies to balance the four bodily humors. For Paracelsus, in fact, the very "book of medicine" was "nature itself" (86), whose plenitude was an invitation to observe and categorize. He justified such efforts to interpret the order of nature in terms of the inherent reason of humankind, which he refers to as the "light of nature" (104). Only humans have the capacity to elicit the potential perfection of nature out through "art," especially the art of alchemy (92–3). Indeed, the central practice of alchemy for Paracelsus illustrates the capacity of humans to make invisible and divine essences visible through art:

> The great virtues that lie hidden in nature would never have been revealed if alchemy had not uncovered them and made them visible. ... the virtues in things remain concealed ... unless the alchemists disclose them, as the summer reveals the nature of the tree ... And each thing has not only one virtue but many, just as a flower has more than one color, and each color has in itself the most diverse hues, and yet they constitute a unity, one thing. Alchemy is a necessary, indispensable art. (144)

Alchemy—the art of reducing the order of nature to its material foundation (*prima materia*) in order to perfect it artificially into gold or the medical potion of *aurum potabile*—reveals the secrets of nature by making them visible and knowable.

The Book and the Body

As in the examples of botany and natural history considered in Chap. 2 above, print technology dramatically reframed medical practice in parts of early modern Europe. The three most popular printed books of *physic* in early modern England were *The Mirror or Glass of Health* (1531), by Thomas Moulton, which appeared in at least nine editions before 1565; an anthology of writings by Galen, Hippocrates, and Avicenna gathered under the title *The Treasury of Health* (1560); and a treatise on uroscopy produced by John Rastell, whose influential printing and playwriting activities we considered in Chap. 3 above. Taken together, these three titles comprise the majority of printed works of medicine in England before 1600.[17] Moulton's *Mirror or Glass of Health* focuses mainly on the plague, at least in the first section, and frames medical treatment in terms of the ancient art of astrology and the more modern faith of Christianity. He founds his study neither on medical authorities nor on actual practice but rather on the four "pillars" of the Church, the saints Augustine, Jerome, Gregory, and Ambrose. Turning to England, Moulton identifies two linked and invisible forces that cause the visible "tokens" of plague: corrupt air caused by the malicious conjunction of the planets Saturn and Jupiter.[18] In addition, those who suffer from "corrupt humors" in their own bodies because of their own sinfulness will be more susceptible to infection by the plague.[19] Rastell's *Here Beginneth the Seeing of Urines* (1525), by contrast, is dedicated to understanding the material signs of illness. The book included over 100 images of blank "piss pots" that could be filled in by users to record the color and consistency of their own urine (Fig. 6.1). Some colors of urine suggest fatal illness, as color "xviii" indicates in the image.[20] Other colors designate health, while still others indicate illnesses that could be treated with herbal remedies.

The role of print in reframing nature was also central to early modern efforts to map the interior of the body through anatomical dissection, especially in the revolutionary work of anatomy by Andreas Vesalius, *Seven Books on the Fabric of the Human Body* (*De humani corporis fabrica libri septem*, 1543). Challenging conventional medical practice, Vesalius framed the human body as a product of the Book of Nature that should be read directly by students of medicine.[21] Vesalius praises anatomical dissection for its capacity to reveal the "marvelous wisdom" of its creator.[22] A rare manuscript about anatomy printed in the sixteenth century that includes a conversation between Vesalius and the famous Paduan

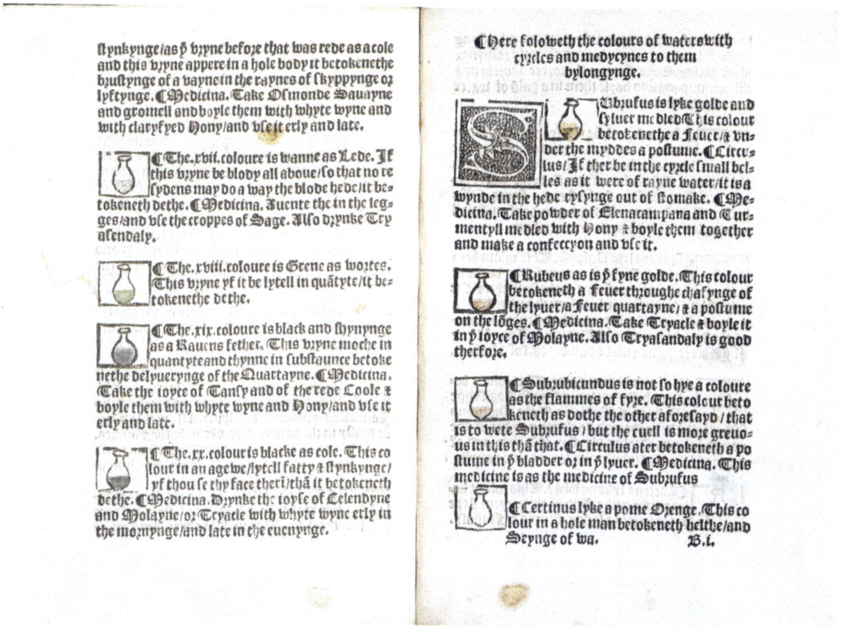

stynkynge/as ẏ uryne before that was rede as a cole and this uryne appere in a hole body it betokenethe brustynge of a vayne in the raynes of skypppynge or lyftynge. ¶Medicina. Take Osmonde Sauayne and gromell and boyle them with whyte wyne and with claryfyed Hony/and use it erly and late.

¶The.xvii.coloure is wanne as Lede. If this uryne be blody all aboue/so that no residens may do a way the blode hede/it betokeneth dethe. ¶Medicina. Aucnte the in the legges/and use the croppes of Sage. Also drynke Tryasendaly.

¶The.xviii.coloure is Grene as wortes. This uryne yf it be lytell in quãtyte/it betokenethe dethe.

¶The.xix.coloure is black and shynynge as a Rauens fether. This uryne moche in quantyte and thynne in substaunce betokenethe delyuerynge of the Quartayne. ¶Medicina. Take the ioyce of Tansy and of the rede Coole & boyle them with whyte wyne and Hony/and use it erly and late.

¶The.xx.colour is blacke as cole. This colour in an agewe/lytell fatty & stynkynge/ yf thou se thy face therĩ/thã it betokeneth dethe. ¶Medicina. Drynke the ioyse of Celendyne and Molayne/or Tryacle with whyte wyne erly in the mornynge/and late in the euenynge.

¶Here foloweth the colours of waters with cyrcles and medycynes to them bylongynge.

Subrufus is lyke golde and syluer medled. This colour betokenethe a Feuer/& under the myddes a postume. ¶Circulus/If ther be in the cyrcle small belles as it were of rayne water/it is a wynde in the hede rysynge out of stomake. ¶Medicina. Take powder of Elenacampana and Turmentyll medled with Hony & boyle them together and make a confeccyon and use it.

¶Rubeus as is ẏ fyne golde. This colour betokeneth a feuer throughe chafynge of the lyuer/a Feuer quartayne/ & a postume on the lõges. ¶Medicina. Take Tryacle & boyle it in ẏ ioyce of Molayne. Also Tryasandaly is good therfore.

¶Subrubicundus is not so hye a coloure as the flammes of fyre. This colour betokeneth as dothe the other aforesayd / that is to wete Subrufus / but the euell is more greuous in this thã that. ¶Circulus ater betokeneth a postume in ẏ bladder or in ẏ lyuer. ¶Medicina. This medicine is as the medicine of Subrufus

¶Certinus lyke a pome Orenge. This colour in a hole man betokeneth helthe/and Seynge of wa. B.i.

Fig. 6.1 From John Rastell, *Here Beginneth the Seeing of Urines* (1525). (Reproduced by kind permission of the Syndics of Cambridge University Library)

professor of medicine Matthaeus Curtius captures this revolutionary approach. Curtius lectured on the textual authority of Galen and Mundinus following the medieval practice in which a lecturer reads from textual authorities while an anatomist performs (see Fig. 6.2). The German manuscript begins by defining anatomy as a "sacred and divine" practice.[23] Distinguishing between the arts of "dissection" (*ars disecandi*) and "anatomy" (*Anathomes*), Curtius argues that particular knowledge of the body from cutting it open should be assimilated into larger understanding of the body's form and structure. Dissection proceeds either by hand (*opera*) or through textual description (*sermone factam*), according to Curtius.[24] But Curtius sides with Galen over Hippocrates about which vein to bleed, causing Vesalius to point to the bodies before them. Confronted by visual evidence, however, Curtius refuses to believe his eyes and challenges Vesalius's hubris in questioning Galen.[25] The constant refrain throughout the lectures celebrating the creative power of

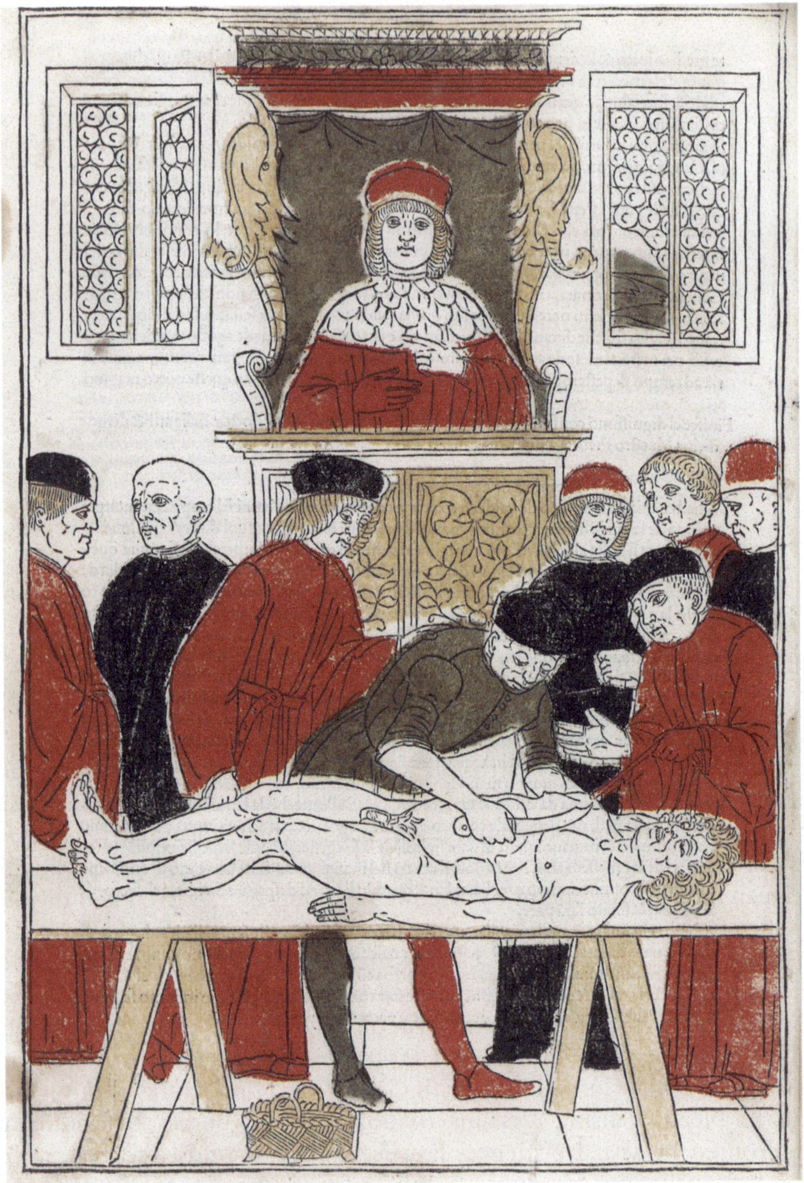

Fig. 6.2 Illuminated woodcut of "The Anatomy Lesson" from the "Pamphlet on Medicine" (*Fasciculo di medicinia*) printed in Venice by the Gregori Brothers in 1494

"Nature" in making the body confirms for Curtius the textual authority of Galen and Mundinus, but for Vesalius such creative power required verification through direct observation and, eventually, through the powerful images and text of the printed *Fabrica* itself.

Even as it became more empirical and oriented toward visual representation, early modern anatomy continued to rely on arts of verbal and physical performance, just as the medical schools remained dependent on forms of oral disputation. As Cynthia Klestinec has demonstrated, these modes of performance foregrounded hearing as well as seeing while seeking to align the study of anatomy with the humanist values of civic honor and fame.[26] Yet the dominance of visualization in the new anatomy theaters of Vesalius and others was also part of a larger process of reinvigorating the Book of Nature and recoding the status of words, images, and objects as forms of knowledge and representation.[27] As David Hillman notes, the *Fabrica* as both a text and a practice within the theater made the invisible body a form of "visible spectacle," though one, as Elizabeth Harvey adds, that also aligns with the power of touch.[28] Given the scarcity of anatomy theaters in early modern Europe and the limited audience within those theaters, the anatomical knowledge produced by the many dissections of Vesalius had its greatest influence in and through moveable type.

Although Vesalius followed Galenic theories of the body in the text of the *Fabrica*, the widely reproduced images of the work were more radical.[29] The 1543 edition of Vesalius's encyclopedia of anatomy incorporated around two-hundred engravings of the tissues, organs, and structure of the human body, most of which were likely produced by Jan Stephen van Calcar, a member of Titian's workshop (Fig. 6.3).[30] This title page illustrates how the famous work of anatomy by Vesalius aligns printing as a technology with modes of viewing the body as a visual text.[31] One reason for the inclusion of so many images in the final version of the book was to provide both knowledge and "pleasure" to those who were squeamish about witnessing dissection in person.[32] In his adaptation of Vesalius for an English audience, the physician and director of Bethlehem Hospital ("Bedlam") in London, Helkiah Crooke, also justifies human dissection on the basis that human bodies are divinely created. God made us stand upright, he argues in *Mikrokosmographia* (1615), so as to "behold and meditate on heavenly things," including our own corporeal forms.[33] Such study of the human body for Crooke therefore produces knowledge not just of the self but also of the "immortal God," who can only be known indirectly through His creations:

Fig. 6.3 Andrea Vesalius, *De humani corporis fabrica* (1543), frontispiece (engraving attributed to Jan Stephen van Calcar)

> So we read in the sacred Scriptures, that *Moses* could not endure the bright shining face of God, his eyes were so dazzled therewith. *The invisible things of God* (sayeth the Apostle) *are known by those things that are visible.* Who is it therefore, that will not honor, reverence, and admire the author and workman of so great a work, if he do attentively advise with himself, how wonderful the fabric and structure of man's body is? *I will praise thee, O Lord,* (saith that Kingly Prophet) *because I am wonderfully made.*[34]

Like the order of nature, the body for Crooke is a visible manifestation that both reveals and conceals its divine origins. Crooke focuses mainly on classical debates over the size, shape, and use of the parts of the body in his work. In the process, however, he takes a less antagonistic position toward Galen and Aristotle than Vesalius, and he foregrounds the body as the unaccommodated "thing itself."[35] Yet, as we have seen throughout this study, the more the order of nature becomes subject to empirical inquiry, the more it reveals itself as both textualized and performative. And the more it reveals itself in this way, the more it requires new forms of explanation to become legible.

Invisible Ailments in *All's Well that Ends Well*

In *All's Well that Ends Well*, Shakespeare alludes directly to the figures of Paracelsus and Galen in a plot that centers on the successful cure of a potentially fatal illness in the King of France by the virginal Helen (or Helena in some versions). To thank Helen for her cure, the recovered king gives her the right to select any man she likes from the French court to be her husband. Her selection, Bertram, pretends to go along with the wishes of the King and Helen before escaping to the wars in Italy, where he hides from his marital obligations and pursues a relationship with an Italian woman named Diana. Shakespeare's source for the story was most likely a tale by Boccaccio that was translated into English as part of the anthology of popular stories collected by William Painter in *The Palace of Pleasure* (1566). A central question in the adaptation of this text is the exact location of the royal fistula. Painter's work locates the royal malady in the heart, and the King in Shakespeare seems to second this location when he notes that his "heart / Will not confess he owes the malady / That doth my life besiege" (2.1.8–10). But, elsewhere in *All's Well*, the Clown's jesting with the Countess about buttocks and his allusion to "French crown" (2.2.16–21) hint that the fistula was located in the King's anus and might therefore be the result of venereal disease.[36] Yet the play never specifies the location of the fistula in the King's body, nor does it reveal the precise method in which Helen restores the King to health.

Such ambiguity mirrors the etiology of the disease itself, whose variability and invisibility made it extremely challenging for both patients and healers. University trained physicians were no more likely to succeed in helping patients suffering from fistulas than empirics, and many remedies suggested by both often make the illness worse.[37] Defined nosologically as ulcers or abscesses opening in the flesh between organs, fistulas could occur in any part of the body, from the tear ducts to legs. In part because of their hiddenness in the body, fistulas also exposed the limits of medical knowledge in early modern England and raised larger questions about the relationship between nature and art. George Puttenham, for example, lists the "fistula *in ano*" in his *Art of English Poesy* as a notorious disease that patients seek to cover up. He refers to it as a "secret disease" that physicians cannot "speedily heal" and that courtiers conceal out of shame.[38] Puttenham urges English poets to leave behind such dissimulation except in employing the "subtleties of his art; that is, when he is most artificial."[39] His discussion of illness, in fact, leads Puttenham to a prolonged examination of the relation between nature and artifice, considering how art can both aid and alter nature in situations such as when a physician can "restore the decayed spirits" of a patient by administering "cordials" or by poets who employ the "metrical proportions" of language to improve nature.[40]

Helen's method of healing also remains ambivalent in the play. She alludes to the "many receipts" that her father, the celebrated doctor Gérard de Narbonne, has given her upon his death, with one in particular designed to cure the King's particular malady (2.1.103). Such "receipts" were part of an ancient vernacular tradition of curing patients that often depended on female knowledge and labor.[41] Gervase Markham, for example, identifies practical knowledge in those healers who have the capacity "to administer any wholesome receipts or medicines" in order to maintain the "health and soundness of the body" and the "preservation and care of the family."[42] In historical context, however, the word "receipts" referred to a Paracelsian rather than Galenic remedy. In act 2, the King aligns himself with the Galenic doctors of the "college" who have given up on his cure. He refuses to "prostitute" his "malady" to mere "empirics" such as Helen (2.1.119–20), employing the word "empiric" in the same way that the Royal College of London did to mean an "unlearned" and even possibly even "fraudulent practitioner of medicine."[43] Helen's successful treatment, by contrast, employs a Paracelsian cure developed in relation to the divine order of nature, illustrating the triumph of unlearned empiricism over university medicine. Helen's miraculous cure signals the victory of the order of nature over the status quo of the order of politics.

Helen claims that her attempt to cure the King reflects "an experiment" of "heaven, not me" (2.1.152) and refers to the "grace" of God (ln. 158) as the instrument of healing. Such religious affirmations also reflect a Paracelsian focus on Christian healing. Humble in her suit to the king, though bold in making it at all, Helen claims to be acting in the name of divine authority: "He that of greatest works is finisher / Oft does them by the weakest minister" (lns. 134–5). Before accepting her offer, moreover, the King observes "some blessèd spirit" operating in Helen (ln. 173), anticipating Lafeu's discernment of the "very hand of heaven" in the cure of the King (2.3.30). Such language frames the cure as a divine miracle, yet in concealing the details of the cure from the audience, it also raises further questions about the relation between medicine, theology, and the *ordo creationis.*

The King notes that his doctors at the "college" of medicine have declared that "laboring art" cannot "ransom nature" (2.1.115–16), suggesting that the illness within him is so deeply implanted by nature that it cannot be treated. For Lafeu, the King's healing by Helen is evidence that miracles have not ceased and that those "philosophical persons" who try to make "modern and familiar things supernatural and causeless" are mistaken (2.3.1–3). Rather than making "trifles of terrors" by reducing them to definite causes, we should "submit ourselves to an unknown fear" (2.3.4–6). Repeating his earlier claim that the King has been revived sexually by Helen, he also frames the royal revival in terms of the order of nature. Lafeu's response to the cure is not to seek specific medical information or to describe the precise procedure used to heal the King, but rather to embrace the supernatural and the unknowability of divinely created nature made visible through embodied performance.

In thanking Helen for her restorative medicine, the King calls her an "immediate heir" to "nature" itself (2.3.130). Paroles confirms that the healing of the King is the "rarest argument of wonder that hath shot out in our latter times" (lns. 6–7), while in the ensuing dialogue Lafeu praises the "artists" who have saved the King, to which Paroles responds, "So I say, both of Galen and Paracelsus" (lns. 9–10). This comic exchange heightens the complexity of the central healing act itself, which is simultaneously clinical and supernatural, virginal and sexualized, comic and tragic. This explicit allusion to Galen and Paracelsus by Paroles is, of course, a red herring. The braggart soldier is trying too hard to fit in at the French court and is eager to appear intelligent before Lafeu. In conflating Galen and Paracelsus, he also reveals his ignorance about the key differences between these two important medical authorities. From another perspective, however, Shakespeare's inclusion of the two most influential medical authorities in Europe at the time—names he mentions nowhere else in his writing—suggests the centrality of medicine to the play. Helen is as both an empirical healer who uses the "art" of medicine in ways that resemble Paracelsus and an embodiment of pure "Nature" whose very presence has the capacity to enliven others. Her secret cure employs theatrical magic to join art and nature in and through an ambiguous act of dramatic representation.

A recurring pattern of concealing and revealing shapes Shakespeare's comedy more broadly. As the play opens, Helen must conceal her true feelings of love for Bertram from those around her. She attempts to recover from the recent loss of her father, whose absence hovers over the play, just as the Duchess mourns her deceased husband and prepares to "lose" her son Bertram to the court of France as a ward of the king (1.1.1). Even Helen's virginity is a negative category of something that is "nothing" and that can be parted with.[44] The extended dialogue between Paroles and Helen about virginity in the opening scene of the play establishes the stakes for this double negative. Helen asks Paroles if there is any "military policy" by which to protect virginity from its greatest enemy, "Man" (lns. 116, 108). Following lewd remarks about virginity being "blown down" by man who will be "blown up" in the process (lns. 118–19), Paroles offers his own windy explanation of how virginity itself violates the "commonwealth of nature":

> It is not politic in the commonwealth of nature to preserve virginity. Loss of virginity is rational increase, and there was never virgin got till virginity first was lost. That you were made of is metal to make virgins. Virginity, by being once lost, may be ten times found; by being ever kept, it is ever lost. 'Tis too cold a companion. Away with't! (lns. 120–26)

This analogy between virginity and the "commonwealth of nature" transforms the order of nature into a political and economic structure to affirm the paradox that virginity must be compromised in order to be supported and sustained. His logic is that of economic reproduction, which

Shakespeare also mobilizes in his sonnets—namely that a virgin woman who loses her virginity through sexual intercourse has the potential to produce many more virgins in the form of her offspring. As Paroles points out, moreover, to speak against virginity is to "accuse your mothers" (lns. 129–30). He then compares virginity to an old cheese that "breeds mites" who eat the cheese (lns. 133–4). Thus virginity, like cheese, "consumes itself" and "dies with feeding his own stomach" (lns. 134–5).

As a comic heroine, of course, Helen is more interested in losing than preserving her virgin state. She has, after all, already confessed her secret love for Bertram. Yet female protagonists in Shakespearean comedy rarely if ever speak so directly about their own virginity. Like the fistula of the King and the secret Paracelsian healing at the center of the play, virginity is not legible to the casual observer. As Shakespeare knew well, moreover, virginity was the subject of important political debate in early modern England, especially in relation to Queen Elizabeth. Even as the queen developed a cult of virginity throughout her adult life as a way of navigating the lineal politics of her monarchy, rumors swirled around court that there was something abnormal about her body. Ben Jonson, for example, asserted that Queen Elizabeth may have had a particularly thick "membrana" covering her vulva that prevented her from losing her virginity and thus from marrying.[45] At stake in the debate was both the vexing question of how the Tudor monarchy could continue into the future and the transgressive nature of female queenship that inspired angst and animosity among early modern authors.[46]

The expansive and unexpected dialogue on virginity in the play's opening scene also pathologizes virginity as a condition of the body that can bring about "green sickness," an illness associated with jaundice and believed to be caused by the buildup of menstrual blood in the female body. This disease was increasingly associated with young, virgin girls in the sixteenth century.[47] Known variously as "chlorosis," the "the virgin's disease," and the "fever of love" (*febris amatoria*), it was typically understood to be an ailment of unmarried women. A common remedy for such an illness, according to many early modern medical manuals, was sexual intercourse. Helen's status as a "green" virgin thus defines both her condition and her path to remedy that condition within the logic of Shakespeare's romantic comedy. Her status as a virgin elevates her healing power with relation to the king, who then allows her to select her own husband in the person of Bertram. The figurative sexual healing of the King, in other words, gives her the leverage to provide her with the actual sexual healing of marriage.

Helen also attempts to conceal her love for Bertram early in the play, making efforts to discern signs of romantic attachment in her parallel those taken to read the body of the King of France for signs of illness, as when the Steward reports to the Countess—who takes on the role of an adopted parent to Helen—that he has overheard Helen confess her love for Bertram. In response, the Countess reflects on her own romantic youth in a series of metaphors of nature:

> Even so it was with me when I was young:
> If ever we are nature's, these are ours. This thorn
> Doth to our rose of youth rightly belong;
> Our blood to us, this to our blood is born:
> It is the show and seal of nature's truth
> Where love's strong passion is impress'd in youth.
> By our remembrances of days foregone,
> Such were our faults, or then we thought them none.
> Her eye is sick on't; I observe her now. (1.3.113–21)

Like other forms of illness, the Countess suggests, love inevitably breaks the "seal of nature's truth" to become legible in the body. Blood, the source of such legibility, remains out of sight and beneath the skin, but it makes its presence known indirectly on the skin through blushing. Helen's natural signification of love only heightens the divine mystery of the central plot of illness, in sharp contrast to the refusal of Bertram to reveal himself to her.

It is significant that both acts of healing remain invisible to the audience. The play moves restlessly between visibility and invisibility, ambivalence that registers in the double negative of Bertram's letter informing Helen he is leaving her: "Till I have no wife, I have nothing in France" (3.2.95). Just as Helen disappears off stage to practice her healing arts with the King, her eventual act of sexual intercourse with Bertram likewise occurs out of sight through the "bed trick" she arranges with Diana. A similar dialectic of presence and absence, visibility and invisibility, shapes the climactic final scene of the play. Bertram has told Helen that he will marry her only if she can show him "a child begotten of thy body that I am father to." In the source text for the play, the Helen figure does in fact produce such offspring, but Shakespeare has Helen refer to herself as pregnant without providing any physical or visual evidence. Bertram and the rest of the court must instead take her word for it, which they do without comment.

Like the opening dialogue on virginity, the concluding scene stages epistemological problems of discernment, including contrasting modes of visual and verbal knowledge and the signifying value of tokens such as rings.[48] The King begins the final scene by declaring that "all is whole" in his forgiveness of Bertram (5.3.37) and in his effort to put the business of Helen behind them: "For we are old, and on our quick'st decrees / Th' inaudible and noiseless foot of time / Steals ere we can effect them" (lns. 40–42). But this attempt to repress the past, which is itself measured against the silent stealth of time's power, unravels, first by the ring that Bertram employs as a "favor" for his new bride, the (absent) daughter of Lafeu, which the King recognizes as belonging to Helen (lns. 81–7). When Bertram cannot provide a satisfactory explanation of how he came into possession of the object, the King, "wrapped in dismal thinkings" (ln. 128), sends him away to prison. Immediately following this disgrace, the King receives a letter from Diana claiming that Bertram has married him and fled Florence without her.

Face to face with Diana, Bertram cannot deny that he does "know" her (5.3.166), but another ring—this time the one that Bertram gave to Diana—produces additional confusion. Diana refuses to answer the King's question about how she found the ring, speaking instead in paradoxes about being both guilty and not guilty and being both a maid and a wife (lns. 283–7). The solution to the cascading "riddle[s]" of Diana, including that the previously dead Helen is now "quick" with child (ln. 297), depends on the physical presence of the "jeweler" herself (ln. 290)—the lost Helen, who "beguiles the truer office" of the King's eyes (ln. 298). Helen's dramatic appearance in the flesh might be read as an effort to cleanse the play of the verbal and visual uncertainties, just as the "presence" of Silvia produces a remedy to idolatrous love at the conclusion of *Two Gentlemen of Verona*. Helen produces the ring that Bertram had demanded from her in order to make their marriage official and claims to be pregnant. In the source text of Painter, the Helen figure produces twin children as evidence of her sexual relationship with Bertram, but in Shakespeare this truth is confirmed only by the word of the mother-to-be. Once again, the body is both revealed and concealed simultaneously. The King, for his part, announces that "all yet seems well" (ln. 326), but his "seems" captures the lingering uncertainty. Bertram echoes such uncertainty in his vow of love: "If she, my liege, can make me know this clearly, / I'll love her dearly, ever dearly" (lns. 309–10) and the King speaks with similar ambiguity in his final lines: "Let us from point to point this story know / To make even truth in pleasure flow" (lns. 318–19). The physical embodiment of signs of truth in the form of the rings and, eventually, of the body of Helen herself does not fully contravene the uncertainties of knowledge that have accumulated over the course of the play.

By turning the pregnant body of Helen into an object of visual and epistemological uncertainty, Shakespeare draws on the joint fascination with and enduring mystery of the female womb in sixteenth-century medical texts.[49] For Paracelsus, pregnancy operated as a sign of invisible divinity: "A child that is being conceived is already a man, although it is not yet visible [. . . .] already resembles the visible man." As he suggests, "in things eternal, faith makes all the works visible; in things corporeal, but not visible, the light of nature makes all things visible [. . . .] Do not judge a thing that can become visible by its present invisibility" (27). At the same time, procreation for Paracelsus must stem from desire, will, and imagination of both men and women. God endows men with reason and desire, but also with the will to determine if and how to use both. Without such will and imagination, no "seed" will be produced in his body. Likewise, women will produce no "seed" without using imagination and free will to focus her desires upon a man (33). As different as they are in form and function—women are beneath men in the great cosmological chain of being, according to Paracelsus—both derive from the same "spirit" in God (36–7).

In contrast to externalized male genitalia, the hiddenness of female sexual organs and modes of reproduction made female sexuality a more frustrating subject of inquiry for the male practitioners of medicine in medieval and early modern Europe. Indeed, as Katharine Park argues, the mysteries of female anatomy came to figure metonymically for the invisible labor of female healing practices.[50] Early modern anatomy epitomized by Vesalius also made the womb more available and more strange simultaneously.[51] The performative body of Helen offers a final structure of visibility and invisibility in the play, in the spirit of the earlier examples of the unseen cure of the king, the bed trick, and the ring trick. The King declares his full knowledge and cognizance of Bertram's past in recognizing the ring itself:

> Plutus himself,
> That knows the tinct and multiplying med'cine,
> Hath not in nature's mystery more science
> Than I have in this ring. (5.3.101–4)

These references to "tinct" and "multiplying medicine" link the King's knowledge of the ring to alchemy and to the "elixir of life" that early modern iatrochemists such as Paracelsus sought as a cure for almost all illnesses.[52] It is humorous that the King would put such faith in this object as a sign of the truth of the relations between Bertram and Helen, especially given widespread skepticism toward alchemy in the period. Some in the audience may well have regarded this passage as a joke, or as an indication of the King's inability to discern truth from falsehood. But, like medical cures that sometimes employed alchemical concoctions, alchemy itself posed epistemological questions about truth and uncertainty in the visible order of nature.[53] The stage prop of the ring becomes, within the logic of the play, an epistemological guarantee of the stable relations between visible and invisible realms.

Mimetic Healing in *Macbeth*

The early modern visibility of natural philosophy at the intersection of print and drama also informs one of the only and most important documents about performances of Shakespearean drama produced during his lifetime—a series of written notes recorded by the notorious astrologer and Paracelsian doctor Simon Forman. Forman gained both fame and the attention of medical authorities by virtue of his skill in casting horoscopes and practising other forms of judicial astrology that were central features of his medical treatments. His clients included the poet Amelia Lanyer and the controversial figure of Frances Howard, who was convicted of having Sir Thomas Overbury murdered.[54] (Forman also bragged about curing his own plague infection in the summer of 1592 by lancing his boils and drinking a potion he concocted.) Imprisoned periodically by local authorities and investigated for his medical practice both by the Barber-Surgeons and the College of Physicians, he nevertheless conducted at least 8000 patient consultations between 1596 and 1603, amassing a small fortune in the process. But the connection between his medical practice and his interest in Shakespeare has received less attention than it deserves.[55]

Forman's medical practice was both secretive and experimental, since it foregrounded trial and error and engaged in the direct observation of nature. In the process, Forman focused on training the senses, especially the eyes, in an effort to discern the legibility of the *ordo creationis*. Forman justified his activities on the basis of God's will, which he considered elemental to all human existence. Such beliefs converge in his poorly received treatise on *The Grounds of the Longitude* (1591), for example. In response to those who doubt his abilities to determine longitude, Forman cites St. Paul in his claim that God's "free grace poured out upon every creature, according to his divine will and foreknowledge" and disparages those who doubt "so bountiful a gift at the hand of God."[56] For Forman, the order of nature invites observers to appreciate its divine origins: "Do not all the creatures of God manifest his power and glory, each one in his kind and nature, as the birds of the air the fishes of the Seas, the beasts and worms that feed on the ground, and the herbs and trees that grow thereof." Just as God showed his "wonders" to Moses, for example, so do plants such as rhubarb and selledony reveal their divine powers of healing.[57]

The fact that Forman recorded his observations about the theater in the same notebook in which he wrote medical notes, horoscopes, and alchemical recipes is not surprising in an age of commonplace books, but the relationship between these topics has not received as much attention as it deserves.[58] His accounts of *Macbeth, The Winter's Tale, Cymbeline* (c. 1609), and *Richard II* recall basic plot points together with advice to himself to "remember" and "observe" what he has witnessed in the playhouse. Writing about *The Winter's Tale,* for example, he urges himself to "Remember also the rogue that came in all tattered like colt pixie"—a reference to the itinerant ballad seller Autolycus, who appears selling deceptive ballads to unwitting consumers in act 4. His summary of observing the trickery of Autolycus is to "Beware of trusting feigned beggars or fawning felons."[59] Likewise, in response to Jack Straw in *Richard II,* who is stabbed at "Smithfield Barrs" because of his lack of street knowledge, Forman admonishes his reader "never admit any party without a bar between, for a man cannot be too wise, nor keep himself too safe" (fol. 201r). But Forman reserves his most extensive comments for *Macbeth,* perhaps because the play aligns more fully than others he witnessed with his own professional activities.

In addition to describing the meeting between Banquo, Macbeth, and the witches (whom Forman calls "women fairies or nymphs"), Forman draws attention to the "many prodigies" that appear in anticipation of the death of Banquo and concludes his account with this note about Lady Macbeth: "Observe also how Macbeth's queen did rise in the night in her sleep and walk and talked and confessed all, and the doctor noted her words" (fol. 207r). It is likely that Forman read as well as saw the tragedy, since he refers to details of the play that were likely not staged in the performance he witnessed. The lesson Simon highlights is perhaps that the guilty will inevitably confess their crimes and that an observant "doctor" like Forman himself will dutifully record the confession. Given Forman's interest in the propensity for all created matter to be part of the revelation of God's will, however, this final sentence by Forman invites special attention. Forman's remarks seem to refer to act 5, scene 1 of *Macbeth,* in which a Doctor and a Gentlewoman observe Lady Macbeth sleepwalking while attempting to wash her hand clean of an imaginary blood stain. The Doctor vows to "set down what comes from her, to satisfy my remembrance the more strongly," as Lady Macbeth wanders at night and tries to remove the invisible stain she sees on her hand in her "Out, damned spot!" monologue (5.1.32–5).

Forman may have taken such note of this episode in the play because Lady Macbeth's symptoms reminded him of those he observed in his own patients. Shakespeare focuses in the scene on problems of medical diagnosis and observation, as well as on larger questions of the status of medicine in relation to divinity and nature.[60] Like Helen in *All's Well,* the Doctor in *Macbeth* has an uncertain relationship with medical practice. He tells Macbeth that he cannot do anything to heal his wife, saying that the "patient" must "minister" to herself, since her condition is one caused by her own conscience (5.3.45–6). But this non-diagnosis does not necessarily set up a sharp divide between medicine and theology in the play. The Doctor also confesses that he does not have adequate evidence to cure her because has not personally witnessed what the Gentlewoman reports—that Lady Macbeth has every night been rising from her bed, putting on her nightgown, unlocking her desk, writing something down, sealing it, and then returning to bed, all while apparently asleep. In the opening lines of the scene, however, the Doctor remarks: "I have two nights watched with you, but can perceive no truth in your report. When was it she last walked?" (5.1.1–3). The Doctor also describes the sleepwalking of Lady Macbeth as a "great perturbation in nature," since she seems to receive the "benefit" of sleep without sleeping (lns. 8–9). For her part, Lady Macbeth's Gentlewoman is unsure of what she has seen and vows not to "report" any words her lady has spoken because just repeating them may get her in trouble (ln. 12). As if on cue, Lady Macbeth enters with a taper, eyes wide open, rubbing her hands as if washing them, then referring to the spot on her hand and noting, enigmatically, "Hell is murky" (ln. 32), to which the Doctor asks, attentively, "Do you mark that?" (ln. 36).

What Forman may have been drawn to, in other words, was not just the strange behavior of Lady Macbeth, who confesses her own complicity in the murder of Duncan in the scene, but also the explicit debate in the play about how to understand medical symptoms. In different ways, both the Doctor and the Gentlewomen understand Lady Macbeth to be admitting more than she knows. When Lady Macbeth asks if her hands will ever be clean, for example, the Doctor chides "Go to, go to. You have known what you should know," to which the Gentlewoman replies, "She has spoke what she should not, I am sure of that. Heaven knows what she has known" (5.1.42–3). Both the Doctor and the Lady come to understand, however, that Lady Macbeth is revealing her own complicity in the murder of Duncan. They also demonstrate how such egregious crimes against nature disrupt structures of representation of speech, witnessing, and communication.

Ultimately, the scene in *Macbeth* to which Forman draws our attention represents the challenges of empirical observation and medical diagnosis, reflecting larger concerns in the play about the sick and insane body.[61] The Doctor and the Gentlewoman engage in their own individual witnessing, yet neither is able to understand Lady Macbeth fully despite her physical presence before them. As the Doctor himself confesses, "this disease is beyond my practice" (5.1.52). Still, as a medical professional, he understands that the "unnatural troubles" he observes may stem from "unnatural deeds" that he refuses to name (lns. 64–5). Like the Gentlewoman, moreover, the Doctor recognizes that Lady Macbeth's actions in the scene ultimately derive from a higher power. As he notes, "more needs she the divine than the physician" (ln. 67). Although the Doctor here seems to divide physic and theology, the scene itself illustrates the marriage of medicine and religion. Lady Macbeth in her madness reveals a truth to which she is herself blind, a truth that depends on an unseen mode divinity working through her that inspires her muddled confession. Her allusion to a murky Hell suggests both that she is imagining herself in that invisible realm and that a particular mode of providence governs the play.

The scene of healing in *Macbeth* thus proposes an artificial model of empirical diagnosis in which the audience participates. It joins other moments of questionable spectacle in the play, including the witches and the dagger that seems to hover in the air before Macbeth just before he murders Duncan.[62] In his recorded observations, Forman emphasizes how *Macbeth* makes the invisible visible in episodes such as the apparition of the dagger and the haunting of the ghost of Banquo in the banquet scene. Forman recalls in particular that Macbeth,

> being at supper with his noblemen, whom he had bid to a feast (to the which also Banquo should have come) [...] began to speak of noble Banquo and to wish that he were there. And as he thus did, standing up to drink a carouse to him, the ghost of Banquo came and sat down in his chair behind him. And he turning about to sit down again saw the ghost of Banquo which fronted him so that he fell into a great passion of fear and fury, uttering many words about his murder, by which when they heard that Banquo was murdered they suspected Macbeth. (fol. 207v)

Here, as in the scenes with Lady Macbeth, a vision unseen by the audience motivates characters to speak against their will and to reveal knowledge that would otherwise remain hidden from others. This unseen force has the appearance of providence, especially as it ultimately punishes the murderers.[63] *Macbeth* thus outlines competing modes of observation, focusing skeptical attention in particular on modes of unmediated vision. Where Forman understood his own medical knowledge, astrological predictions, and alchemical cures to be expressions of divine will, the Doctor in *Macbeth* turns to God only to register his inability to treat her. For both, however, acts of medical observation occur within a theological framework.

The Doctor complains that "My mind she has mated [i.e., baffled] and amazed my sight" (5.1.71). Unable to process the vision, he takes refuge in the written word of his notes. At the same time, the attending Gentlewoman takes pains to conceal from the world what she has witnessed but not understood, noting that the strange vision of Lady Macbeth can only be known through ineffable "Heaven" (ln. 43). As audience members, we are aware of Lady Macbeth's role in the regicide at the center of the play, but there are limits to how we understand her motivations. The words and actions of Lady Macbeth are no longer her own, and we do not achieve clear knowledge of the forces speaking through her. The Doctor's humble recognition of the limits of his own ability to understand what he sees contrasts with Macbeth, who places enormous faith in the prophetic visions of the three witches and for whom the hallucinated dagger becomes a sign of his necessary act of killing. Shakespeare stages a conflict between the two figures in 5.3 when Macbeth asks the Doctor about the health of his wife:

MACBETH How does your patient, Doctor?
DOCTOR Not so sick, my lord,
As she is troubled with thick-coming fancies
That keep her from her rest.

MACBETH	Cure her of that. Canst thou not minister to a mind diseased, Pluck from the memory a rooted sorrow, Raze out the written troubles of the brain, And with some sweet oblivious antidote Cleanse the stuffed bosom of that perilous stuff Which weighs upon the heart?
DOCTOR	Therein the patient Must minister to himself.
MACBETH	Throw physic to the dogs, I'll none of it.— Come, put mine armor on; give me my staff. (5.3.37–48)

In response to his wife's insomnia caused by "fancies" in her head, Macbeth generates his own macabre fantasy of "written troubles" that produce a "rooted sorrow" in Lady Macbeth, troubles he hopes to eliminate with a "sweet oblivious antidote." Yet, the Doctor notes, such a cure can only come from the patient herself, causing Macbeth to reject the potential healing power of professional medicine and turn toward the battle at Dunsinane.

Lady Macbeth's malady not only stumps the medical professionals who attempt to treat her—it also figures the limitations of medical knowledge in early modern Europe. Simon Forman's intense scrutiny of the play may emerge from this perceived critique of his central profession. Shakespeare also links skepticism toward medicine in to broader skepticism toward the capacity of human vision to produce knowledge in the play. The play includes and questions spectacular visions, including Macbeth's hallucination of a dagger with a handle pointed toward him and the three apparitions produced by the witches in act 4, including the "show of eight kings" with Banquo at the last holding a mirror in his hands (4.1.111–24). This latter vision probably relied on a trick mirror that would have allowed Macbeth and the audience to see the face of King James in its reflection.[64] Such a sophisticated stage device suggests the necessary mediation of mimesis as well as the importance of the material conditions of staging and viewing drama that shaped the experience of Forman and others who witnessed live performances of Shakespeare's tragedy in early modern London.

Notes

1. Nancy G. Siraisi, "Medicine, 1450–1620, and the History of Science," 492.
2. For overviews of early modern medicine with special attention to the role of religion, see Siraisi, *Medieval and Early Renaissance Medicine*, esp. 2, 7–10, 14, 38–9, 43–7, 115–19, and *The Healing Arts: Health, Disease, and*

Society in Europe, 1500–1800, ed. Elmer, esp. xvii, xix–xx, 18–21, 84–106, 148–50, 154.

3. Thomas Gale, "A Brief Declaration of the Worthy Art of Medicine," in *Certain Works of Galen*, sig. A1. See also Kocher, *Science and Religion in Elizabethan England*, 260.
4. John Harley Warner, "The History of Science and the Sciences of Medicine," esp. 188.
5. As Andrew Weeks suggests, a "theological pattern grounds the thought of Paracelsus as much as that of Luther." See *Paracelsus: Speculative Theory and the Crisis of the Early Reformation*, 128.
6. Galen, *Certain Works*, sigs. A1-A1v . Paracelsus, *Astronomia magna* (1571), in *Paracelsus: Selected Writings*, ed. Jacobi, 195–6. All references to Paracelsus are to this edition, which is hereafter cited parenthetically in the text by page number.
7. Baxter described Paracelsus as a "Drunken Conjuror, who had converse with the Devils," while Hall dismissed him as a "Libertine" and a "drunkard" who had the "Devil for his pursebearer." Peter Elmer includes both quotes in his account of "Medicine, Religion, and the Puritan Revolution," in *The Medical Revolution of the Seventeenth Century*, ed. French and Wear, 30.
8. Allen G. Debus analyzes the contrast between Galen and Paracelsus in early modern Europe. See, e.g., *The Chemical Philosophy*, 51–61 and *The English Paracelsians*, esp. 56–88 and 66. See also the overview of both Galenic and Paracelisan medicine in F. David Hoeniger, *Medicine and Shakespeare in the English Renaissance*, 71–130, esp. 123. For a more recent study, see Bruce Moran, *Paracelsus: An Alchemical Life.*
9. On the status of female healers in particular, see Leigh Whaley, *Women and the Practice of Medical Care in Early Modern Europe*, 48–90.
10. This early by-law of the Royal College of Physicians in London is numbered CLXVII. For discussion and citation, see C.J. Sisson, "Shakespeare's Helena and Dr. William Harvey," 7. See also Hoeniger, *Medicine and Shakespeare in the English Renaissance*, 298.
11. Deborah E. Harkness, *The Jewel House*, 60–61. Debus also emphasizes practical "compromise" between the two "camps" of Galen and Paracelsus for many doctors in "Guinthrius, Libavius and Sennert ," in *Science, Medicine, and Society in the Renaissance*, ed. Debus, 151. Elsewhere, Debus notes that most printed works of medicine reveal indifference rather than hostility toward iatrochemistry in sixteenth-century England and that the Royal College of Physicians in London included chemical remedies in a pharmacopoeia that was planned for publication in the 1580s. See *The English Paracelsians*, 76–77.
12. See Julie Orlemanski, *Symptomatic Subjects*, 58–9.

13. Galen, *Certain Works*, sig. B7.
14. Ibid., sig. C4v.
15. See Andrew Cunningham, *The Anatomical Renaissance*, 244.
16. See Walter Pagel, *Paracelsus*, 83.
17. These statistics derive from my own survey of printed books of medicine in the *Early English Books Online* database.
18. Thomas Moulton, *The Mirror or Glass of Health*, sigs. A4v – B2.
19. Ibid., sig. B3v.
20. Anon., *Here Beginneth the Seeing of Urines*, sigs. A4v-B1
21. On the body as a text made by God to be read by medical students, see Ole Peter Grell, "Medicine and Religion in Sixteenth-Century Europe," in *The Healing Arts*, ed. Elmer, 87.
22. Galen, *On the Usefulness of the Parts of the Body*, 68.
23. The account by Baldasar Heseler has been edited and translated into English in *Andreas Vesalius' First Public Anatomy at Bologna, 1540*, ed. Eriksson, 44–5.
24. Ibid., 46–7.
25. Ibid., 272–3.
26. Cynthia Klestinec, "Civility, Comportment, and Anatomy Theater," 435–6.
27. Klestinec, *Theaters of Anatomy*, 6–7.
28. David Hillman, "Visceral Knowledge," in *The Body in Parts*, ed. Hillman and Mazzio, 83. See Elizabeth D. Harvey's extension of this argument in "The Touching Organ," in *Sensible Flesh*, ed. Harvey, esp. 82–3.
29. See, e.g., Debus, *Man and Nature in the Renaissance*, 60–63.
30. Kusukawa, *Picturing the Book of Nature*, 185–90. As she notes, it is also possible that Titian himself contributed images to the book.
31. Daston, "Observing Nature," in *Prints and the Pursuit of Knowledge in Early Modern Europe*, ed. Dackerman, 127.
32. Vesalius, *De humani corporis fabrica*, 1:lvi. See also Kusukawa, *Picturing the Book of Nature*, 213.
33. Crooke, *Mikrokosmographia*, 5.
34. Ibid., 14.
35. On the phrase the "thing itself," see Jonathan Sawday, *The Body Emblazoned*, 11–2, who draws on an argument made by Linda Williams about hardcore pornography as a version of the "thing itself" (*Hard Core*, 48–9).
36. Despite the fact that the King himself locates the "malady" of the fistula near his heart, many critics continue to argue that he suffers from a *fistula in ano*. See, e.g., Catherine Field, "Sweet Practicer, thy Physic I will Try"; Bruce Boehrer, "The Privy and its Double," in Dutton and Howard, eds., 4:83–4; and Cosman.
37. See Luke Demaitre's discussion of the fistula in *Medieval Medicine*, 95–6.
38. Puttenham, *The Art of English Poesy*, ed. Whigham and Rebhorn, 380.
39. Ibid., 382.

40. Ibid., 383–4.
41. For fuller accounts of female healing, see Sharon T. Strocchia, *Forgotten Healers*, and Susan H. Brandt, *Women Healers.*
42. Gervase Markham, *The English Housewife*, ed. Best, 8.
43. For example, Hilarius the gardener refers to "empirics" as "unlearned practitioners" in William Bullein's *Bulleins Bulwarke*, fol. 70v. See the *Oxford English Dictionary*, s.v., "empiric," def. 2c.
44. See William C. Carroll, "The Virgin Not," in *Shakespeare and Gender*, ed. Kamps and Barker, 283–301.
45. As Kaara Peterson has noted, for example, the postmortem body of the queen underwent extensive surveillance to discern her virgin status. See "Elizabeth I's Virginity and the Body of Evidence."
46. See, e.g., the essays in *Dissing Elizabeth: Negative Representations of Gloriana*, ed. Walker; and Katherine Eggert, *Showing Like a Queen.*
47. See Helen King, *The Disease of Virgins*, esp. 27–29, 41–63.
48. On the sexual economy of the ring trick in *All's Well*, see Kaara Peterson, "The Ring's the Thing."
49. See, e.g., Katherine Eggert's chapter on "How to Avoid Gynecology" in *Disknowledge*, 157–205.
50. See Park, *Secrets of Women: Gender, Generation, and the Origins of Human Dissection.*
51. For more on the dissection of the womb, see Park, "Dissecting the Female Body."
52. On alchemical medicine (iatrochemistry), see Allen Debus, *The Chemical Philosophy*, 21 et. passim, and Mary Lindeman, *Medicine and Society in Early Modern Europe*, esp. 102–7.
53. For more on the relation between alchemy and medicine in early modern Europe, especially in regard to Paracelsus, see the essays gathered in *Bridging Traditions*, ed. Parshall, Walton, and Moran.
54. Howard's letter to Forman asking for help in defeating the affections of her husband, the Earl of Essex, and increasing the desire of her lover, the Earl of Somerset, was uncovered in her murder trial. For historical accounts of the relation between Howard and Forman, see Alastair Bellany, *The Politics of Court Scandal in Early Modern England*, 95–6 and Barbara Howard Traister, *The Notorious Astrological Physician of London*, 181–3.
55. S.P. Cerasano does consider some of Forman's medical and occult interests in relation to the culture of early modern theater in "Philip Henslowe, Simon Forman, and the Theatrical Community of the 1590s."
56. Forman, *The Grounds of the Longitude,* sig. A3.
57. Ibid., sig. A3.
58. Thus Lauren Kassell makes almost no reference to Forman's observations about Shakespeare in her study of *Medicine and Magic in Elizabethan London.* In literary analysis, Taryna Cooper makes no connection between Forman's medical practices and his interest in Shakespeare in her discussion of Forman in *Searching for Shakespeare*, 148.

59. The manuscript containing Forman's observations is Bodleian MS Ashmole 208, fol. 200–207b. A useful digital edition produced by the Folger Shakespeare Library can be found at https://shakespearedocumented.folger.edu/resource/document/formans-account-seeing-plays-globe-macbeth-cymbeline-winters-tale. All quotations from Forman are from this digital version, hereafter cited parenthetically in the text by original folio signature.
60. As Kocher notes in his analysis of the role of the doctor in the play, Shakespeare addresses "one of the chief medico-religious questions of his age" in portraying the melancholia of Lady Macbeth. But he also draws inaccurate distinctions between medicine and religion—as well as between materialism and Christianity—in his reading of Lady Macbeth's self-willed melancholia as an "escape from conscience." See "Lady Macbeth and the Doctor," 348–9.
61. Suparna Roychoudhury puts it well when she notes that the play expresses "anxiety over the border between the interior and the exterior, the native and the alien, sanity and sickness." As she notes, the play reveals "the fine line between hallucinatory fantasy and ordinary subjectivity; the difficulty of diagnosing mental diseases as such, and the limitations of the conception of personhood as contained interiority." See "Melancholy, Ecstasy, Phantasma," 205.
62. For a fascinating study of the dagger in relation to early modern debates about vision, see Stuart Clarke, *Vanities of the Eye*, 254–6. For a broader account of the visual economy of the play, see Diehl, "Horrid Image, Sorry Sight, Fatal Vision."
63. There is a similar pattern in the anonymous domestic tragedy of *Arden of Faversham*, though in that case providential force of correction appears in more tangible forms in the snowy prints that reveal the moving of the body of Arden and ultimately lead to the confessions of Alice and her fellow murderers.
64. Marjorie Garber, *Shakespeare's Ghost Writers*, 115–118.

CHAPTER 7

The Nature of Divination in *Julius Caesar* and *Antony and Cleopatra*

Beginning in November of 1577, millions of people around the earth noticed a strange light in the sky that remained visible both day and night to the naked eye. The fiery object blazed in the firmament for several months, eventually becoming known as the "Great Comet." By the time it disappeared from earthly view in March of 1578, it had become one of the most widely documented astrological events of the century.[1] One especially close observer of the celestial object was the Danish astronomer and mathematician Tycho Brahe, who also made some startling calculations. He estimated that the comet traveled at an altitude of at least 50,000 German miles (c. 379,500 km), placing it well beyond the sphere of the moon. At such a distance, Brahe noted, it challenged Aristotle's theory that superlunary objects remained "fixed" in concentric spheres.[2] Brahe's remarkably precise measurements of the Great Comet, together with his studies of the "nova" of 1572 and other stellar events in the 1580s, posed a fundamental challenge to Aristotelian cosmology and helped to elevate mathematics and empirical observation as essential tools of astronomy for future students of the cosmos.[3]

But Brahe's empirical interest in the Great Comet was not purely empirical. As he wrote in his treatise on the comet, he was motivated by a desire to "contemplate" God's "almighty and indescribable marvel of the heavens, investigate it by means of the discernment which He has given

A. Kitch, *Shakespeare's Theater of Nature*, Palgrave Studies in Literature, Science and Medicine,
https://doi.org/10.1007/978-3-031-78082-0_7

us, and thus also learn to know rightly the Creator of it."[4] Like many fellow natural philosophers, in other words, Brahe viewed the comet as both a source of spiritual insight and a product of the *ordo creationis*. Brahe expressed his humility as a reader of the heavens, noting that "we humans, with our limited and earthly understanding, really have no proper grounds or conceptual framework for explaining the material of comets and how they are generated." His theology, is also apophatic in his observation that the divine creator who produced the Great Comet remained both "unfathomable and unknown."[5]

Brahe's response to the Great Comet was not unusual among his fellow astronomers. Copernicus, for example, had referred to the stars thirty years earlier as the product of "the best and most orderly workman [i.e., God]" in his pathbreaking study *On the Revolutions of the Heavenly Spheres* (*De revolutionibus orbium coelestium*, 1543).[6] Indeed, the study of the stars was fully embedded in theological assumptions about the *ordo creationis* and the prophetic capacity of the divine signs it contained. Classical, medieval, and early modern astronomy also included forms of divination. Examples abound from ancient Egypt, Greece, and Rome of diviners studying the order of nature for information about the future—information that depended on foundational assumptions about God's active role in the universe. Such divination helped to generate what Peter Struck calls "surplus knowledge" that was simultaneously intuitive and rational.[7] Like his precursors, Brahe claimed that the Great Comet displayed God's anger and foretold ensuing storms, droughts, earthquakes, spoiled harvests, pandemics, and "great disunity among reigning potentates" in Europe.[8]

Brahe's response to the comet merged older and newer epistemologies in the service of scientific discovery. His explanation of the Great Comet followed ancient traditions of divination that attempted to discern the order of nature for divine truths while also responding to debates within more recent Christian traditions about divination as an art and science. Many Christians disagreed about the value of divination. Although some gnostic sects appropriated Roman divination in the early centuries of Christianity, most patristic authors tried to suppress divination. Both Origen and Augustine, for example, worked to restrict divination to the canonical forms of prophecy recounted in the Old Testament and the clearly demarcated signs of the apocalypse in Revelation. Augustine also argued that the only acceptable medium between man and God was Christ. He therefore relegated artificial divination to a lowly discipline that allows for skills to be cultivated but has no virtue or sense of purpose. At the same time, perhaps influenced by

Stoicism, Augustine embraced divine providence.[9] Although many Catholic theologians followed Augustine in condemning divination as a form of pagan magic, some humanists helped to revive interest in the ancient art in early modern Europe. The Florentine translator of Plato and of the *Corpus Hermeticum*, Marsilio Ficino, for example, defended divination as a form of ancient wisdom, a *prisca theologia* that anticipated Christianity. In his translation of the hermetic text of *Pimander* completed in 1463, Ficino calls Hermes an "ancient Heathen" who was "guided ... by a mighty sense of the interior Excellency of the Soul of Man." Even though he was guilty of "some Errors," Hermes still bore witness to the truth of Christianity.[10]

Protestants writing in early modern Europe raised a series of particular and pressing questions about divination and the art of astrology more broadly.[11] Martin Luther, for example, warned in a sermon on Luke that God knows better than "any Soothsayer or Astronomer" how to read and interpret the signs of apocalypse. Only the bible as the word of God explains these signs accurately, for Luther, with the apophatic caveat that His will is ultimately inscrutable.[12] Luther's follower Philip Melanchthon was more sympathetic to astrology and other modes of divination, even taking part in a debate about Luther's own horoscope with the Italian astrologer Lucas Gauricus.[13] Yet the science of astrology "was no longer a default mode of approaching futurity," as Carla Mazzio notes, even if its popularity was confirmed by the hundreds of thousands of almanacs printed year after year in England and elsewhere in Europe.[14]

For English natural philosophers, the art of reading the stars was likewise subject to new scrutiny. Francis Bacon, for example, dismissed astrology as a form of "superstition," though he hoped to see the ancient practice "purged" rather than rejected.[15] Although he admitted that the movement of the stars could influence both humans and the natural world, Bacon urged natural philosophers to conduct experiments in order to evaluate the efficacy of past predictions. He found greater validity, moreover, in reading the order of nature for prophetic signs than he did in electional astrology, in which astrologers predicted the best times for certain activities, and the casting of individual horoscopes.[16] Astrology for Bacon, in other words, was legitimate to the degree that it represented an older form of the *venatio* or "hunt" for knowledge on which he hoped to ground a new model of inductive philosophy.

This chapter situates two of Shakespeare's Roman plays, *Julius Caesar* and *Antony and Cleopatra*, in relation to classical, medieval, and early modern debates about divination and to larger questions about the status of "nature" in early modern Europe. I suggest that divination in both plays becomes a contested mode of knowledge that reveals philosophical differences between characters and helps more broadly to recast Roman political history within the universalist theology of a Christianity. Astrological speculation also shapes the important racial dynamics of *Antony and Cleopatra,* as I explain more fully below. Shakespeare's foregrounding of divination in both plays acknowledges the historical Caesar's own role in the history of divination, especially since his death was understood as a disruption to the order of nature. The Roman historian Gaius Suetonius, writing during the Roman Empire, includes in his account of Caesar's life and death several allusions to the "unmistakable signs" of such disruption leading up to his assassination. For example, he details the discovery of a tablet in a newly disturbed grave that predicted Caesar's death and the weeping of horses that he had used to cross the Rubicon. Suetonius also described what would become the most famous supernatural event associated with Caesar—a comet that shone for "seven successive days" during his funeral games and that was "believed to be the soul of Caesar, who had been taken to heaven."[17] This story followed Pliny, who provided an account of Caesar's translation to the stars ("catasterism") in his *Natural History* and noted how unusual it was for Romans to regard comets as anything other than portents of evil.[18]

Shakespeare's central source for his Roman plays, Plutarch's *Lives of the Noble Grecians and Romans* (c.100 CE), likewise includes the comet as one of many "strange and wonderful signs" at Caesar's death—along with a flaming sky, nocturnal birds appearing during the day, "spirits running up and down in the night," and various body parts engulfed by flame.[19] Another favorite source of Shakespeare, Ovid's *Metamorphoses,* also describes the comet or "blazing star" as evidence of Caesar's lasting fame and triumph.[20] Shakespeare's Roman plays invoke this tradition by staging moments of both "natural" divination in which signs of divination appear unexpectedly to observers of the order of nature and "artificial" divination in which augurers and haruspex seek such signs through rituals such as sacrifice and the staged observation of animals kept for the purpose.[21] Both forms of divination foregrounded questions of empirical observation of the *ordo creationis* as well as relations between artificial and natural representation.

Even though Shakespeare does not stage the famous comet associated with Julius Caesar directly, his tragedy contains more instances of divination than any other play he wrote.[22] *Julius Caesar* invokes divination in retelling Caesar's death, including the Soothsayer's warning to Caesar to "beware the Ides of March" (1.2.18) and the strange portents that appear during the storm scene in 1.3. Shakespeare links such scenes to central questions about politics and religion in the play.[23] In one sense, divination rushes in where certain knowledge fails, inviting the audience to employ its privileged knowledge about the future that is mostly invisible to the characters themselves. The soothsayer and the augurers in *Julius Caesar* are correct in their prophecies about the danger that Caesar faces in Rome, but only up to a point. These diviners require supplemental knowledge that can only be supplied by a Christian future that they cannot see, namely the emergence of the Roman Empire as a nursery for the birth of Christianity.

Both *Julius Caesar* and *Antony and Cleopatra* also depict, albeit subtly, the political ascent of Octavius Caesar—the adopted son of Julius Caesar—who became the first Emperor of Rome. London audiences witnessing original productions of the plays were far more likely than their modern contemporaries to register that Caesar's death also ended the age of the Republic, ushering in not just the era of Empire but also of Christianity, since the Augustan Empire established the "universal peace" in which Christ was born (*Antony and Cleopatra* 4.6.5). Divination thus helps Shakespeare to suture a pagan past to a Christian present and future, while inviting the audience to consider its own place in the history of divination. Modes of divination in both *Julius Caesar* and *Antony and Cleopatra* also generate hermeneutic ambivalence of the kind that Brahe and other astronomers invoked in their works of natural philosophy, raising larger questions about how to read the *ordo creationis*. By expanding and elevating divination in his Roman plays, Shakespeare also makes it difficult for any single character to recognize his own position in relation to the divine order of nature.[24] Like other plays we have analyzed, moreover, Shakespeare's Roman plays reframe the order of nature in terms of mimesis, establishing theater as an *ordo artificialis* through which to understand the *ordo creationis*.

Tokens of the Gods

Shakespeare was clearly familiar with the classical accounts of the comet that blazed following Caesar's death, though, following many early modern versions of the story, he reverses the chronology of events. In *Hamlet*, for example, Horatio alludes to the "stars with trains of fire and dews of blood" that marked the sky "a little ere the mightiest Julius fell" (1.1.113, 116). Shakespeare here follows the early modern practice of interpreting the comet not as a sign of Caesar's catasterism *after* death, but rather as a warning of his impending demise. In a study of comets translated into English in 1577, for example, the German astronomer Frederic Nausea asks, rhetorically, "Who is ignorant, that after the appearance of a Comet in the time of *Julius Caesar,* not only civil wars followed, in manner most lamentable, but also the death & murdering of *Julius Caesar* his own person, which was most miserable?"[25] Shakespeare does not stage the comet directly in *Julius Caesar*, even though Calphurnia does note that "when beggars die there are no comets seen" and that the "heavens themselves blaze forth the death of princes" (2.2.30–1). But he does stage a storm scene in act 1 that raises question of divination through its extraordinary images of a "tempest dropping fire," lions walking the streets, bodies engulfed in flame, and nocturnal owls hooting at midday (1.3.10–27). Casca names these visions "prodigies" and reads them as "tokens" and "heralds" sent from the "most mighty gods" to "astonish" Romans (lns. 28, 55–56). But Cassius offers a different reading of these "portentous things" (ln. 31):

> You are dull, Casca, and those sparks of life
> That should be in a Roman you do want,
> Or else you use not. You look pale, and gaze,
> And put on fear, and cast yourself in wonder
> To see the strange impatience of the heavens.
> But if you would consider the true cause [...]
> Why all these things change from their ordinance,
> Their natures, and pre-formèd faculties,
> To monstrous quality—why, you shall find
> That heaven hath infused them with these spirits
> To make them instruments of fear and warning
> Unto some monstrous state. (1.3.57–62, 66–71)

Assuming the mantle of a natural philosopher, Cassius seeks the "true cause" of the disorders of nature, drawing on the Latin root of the word "monster," *monstrare*, as "to demonstrate" or make visible. To make the state "monstrous" is to make it a subject of visual inquiry that seeks to uncover laws and norms. It also makes the political crisis unfolding in Rome into an object of divination and natural philosophy. Ludwig Lavater's *Ghosts and Spirits Walking in the Night* (1572) catalogs similarly monstrous visions that violate norms of nature while also giving "warning of things to come." [26] Citing the "well-known" story of the "signs that went before the death of *Julius Caesar*" (164), Lavater invokes the debate between the Epicurean Cassius and the Stoic Brutus over whether certain "visions" can be attributed to natural causes (9). The strongly Protestant Lavater, who was the son-in-law of the Swiss reformer Heinrich Bullinger, acknowledges that some visions can be entirely generated by melancholic minds, but defends the reality of many visions of nature, all of which—even those that are Satanic—originate in God. At stake in the debate was how to understand supernatural events in relation to divine providence.

For Cassius, the monstrous transformation of nature from its "preformèd faculties" offers a kind of political theology. Resisting Casca's interpretation of the storm, he urges action, seeking to align the "wonder" of the storm with his own campaign against Caesar. Cassius's "monstrous state" also reframes the classical story of Caesar's fall in relation to futures that characters in the play cannot anticipate, including Christian debates about interpreting the order of nature. For Augustine, no portent or prodigy could be "against nature" (*contra naturam*), since all things in nature have been created by God. What we take to be monstrous derives in truth from our ignorance of God's plan for the universe. But for later authors, including Isidore of Seville, portents, omens, monsters, and prodigies "show [*monstrare*] and predict future things," since "God wishes to signify the future through the faults that are born [...] by which he forewarns and signifies to peoples or individuals a misfortune to come."[27] God, in other words, uses natural and supernatural signs as a way to warn Christians about impending social and political disorder.

The "monstrous state" described by Cassius was quite familiar to those Protestants in Europe who sought to justify their critiques of Catholicism by way of portents and prodigies in the natural world. Early Modern England was itself awash with prophetic texts, including adaptations of the Greek and Roman Sibylline prophecies that described the Pope as the Antichrist and pamphlets anticipating the Second Coming. John Foxe incorporated apocalyptic writings from a book of Sibylline prophecies in the 1570 edition of *Acts and Monuments*, while authors from the translator Stephen Batman to the choral composer John Merbecke utilized Sibylline imagery to advance Protestant nationalism.[28] *The History of Strange Wonders* (1561) by the German Protestant and follower of Melanchthon Joachim Camerarius likewise interprets a number of strange sights around Europe in terms of the Second Coming and the struggle between Protestants and the Roman "Antichrist."[29] Against naturalistic explanations of prodigies by Epicurus and his followers, Camerarius insists that the strange wonders are "above nature" and have been sent by God to warn of future calamities:

> Therefore these things showing themselves so often times, and so wonderful, verily we ought to understand and to think, that the moving and working of nature is wrested out of frame, and the state and condition of men to be turned out of course, and that the effects of nature being weak and sick may no longer continue nor endure: for even as monstrous births do not live long, so likewise the degenerate and monstrous state of this corrupt world shall not last long.[30]

Prodigies may disrupt the order of nature, but they also confirm the divinity within and behind that order. The "degenerate and monstrous state" of the corrupt secular world is only temporary in relation to the divine providence that orchestrates its teleological course.

Shakespeare's representation of divination in his Roman plays also draws deeply on Plutarch. A priest at Delphi for the last thirty years of his life, Plutarch frequently praised modes of divination in his *Lives* of famous Greek and Roman men. To remedy what he saw as the vulgarity of his sources in Thucydides, Nepos, and Suetonius, Plutarch explicitly incorporated accounts of portents in his biographical studies in order to demonstrate the power of divination and provide a moral framework for human existence.[31] In his retelling of the death of Caesar, however, Plutarch downplays the lightning and thunder heard on the night before the murder of Caesar, as well as the appearance of some "wild birds" that do not appear in Shakespeare (890). The lack of precision in the portents, as well as in the accounting of them by Plutarch, only heightens the uncertainty of the moment.

For both Plutarch and Shakespeare, in other words, divination entails epistemological and hermeneutic problems about how to interpret nature and human action. Both also link divination to questions of natural philosophy. In his life of Pericles, for example, Plutarch describes a tension between natural philosophy and prognostication in relating the competing interpretations offered by the natural philosopher Anaxagoras and the diviner Lampon over a strange object that Pericles once acquired: a ram's head with one horn in its middle. Lampon interprets the omen as a sign that the government of Athens, which at the time was divided between followers of Thucydides and Pericles, should unify around Pericles, but Anaxagoras, after cutting into the skull to examine the irregular brain of the ram, dismisses this interpretation as mere superstition. Plutarch notes, however, that both could be right, since the natural philosopher determines the natural "cause" of the object while the diviner provides an account of its "end" or purpose (170). Plutarch thus takes a skeptical but balanced approach toward divination in his *Lives*, echoing the influential treatises on divination by Marcus Cicero.

Cicero's treatises *On Divination* (*De divinatione*, c. 44 BCE) and *On the Nature of the Gods* (*De natura deorum*, c. 45 BCE)—both completed in the aftermath of Caesar's assassination—frame divination in ways that also resonate with Shakespeare's Roman plays.[32] *On the Nature of the Gods* features the Stoic philosopher Lucilius Balbus, who sets out to prove not just the existence of the gods, but also their character and ability to shape human action. For Balbus, divination demonstrates that the Gods look out for the best interest of humans by giving them advance knowledge of the future. "The Gods," he argues, "take care of great things and disregard the small," while for "truly great men" all "prosperity" is given (105). His interlocutor, the Epicurean senator Cotta Velleius, does not reject the existence of the gods, but challenges the attribution of divinity to even the smallest creature on earth (118) as well as the implication that "some Deity appoints and directs" the tide (113). For Epicureans such as Cotta, the numerous gods and their attributes celebrated by Balbus are only so many "stories" that have accumulated over time to produce a complex and illogical set of "fables" (126–7). Cicero expands on this debate about both the agency of divinity in the *ordo creationis* and the capacity of mankind to learn about that agency in his dialogue *On Divination*, where Quintus mounts a defense of the gods similar to that of Balbus, while Marcus adds to the skeptical claims of Cotta. For Quintus, both "natural" divination (e.g., dreams, meteorological events, and supernatural visions) and "artificial" divination (e.g., the interpretation of birds or sacrificial animals by augurs, haruspex, and other diviners) demonstrate that "everything is filled and pervaded by a divine intelligence and eternal sense," suggesting that the soul of humankind is a "kindred" soul with God who, in turn, produces signs of the future (190).

Examples of this signaling for Quintus include the "prodigies" surrounding Caesar's death, such as the absent heart in the sacrificial ox discovered by Caesar's augurers (193). Against Stoic claims by Quintus defending divine intervention in human affairs, Marcus suggests that, even if Caesar had known the future, he would not have wanted to, since had he foreseen his own death, "in what wretchedness would he have passed his life?" (209). Marcus thus identifies a paradox at the heart of his brother's defense of divination, namely that Stoics regard everything in the universe as both predetermined by fate and subject to chance:

> If we cannot foresee anything which happens by chance, since that thing is necessarily uncertain, therefore there is no divination; and if, on the contrary, things are to happen can be foreseen because they happen by an infallible fatality, there is no divination, because you say that divination only relates to fortuitous events. (210)

A related paradox for Marcus is that divination apparently reveals universal *sympathy* between earth and the heavens, even though there is no inherent connection between, for example, the "cleft of a liver" and the "universal nature of things" (213). Not only haruspicy, but also augury and the interpretation of dreams make false assertions about the relation between universal and particular. Divination for Marcus avoids the real motivator of human action—virtue—by ascribing such action to the gods (233). The only solution, Marcus ultimately claims, is to "reject" and "eradicate" all forms of divination as "superstitious errors" (262).

Cicero's skepticism toward divination did not keep him from practicing it, however. He was himself a member of the Roman College of Augurs, though he noted in a letter to Aulus Caecina that he did not need to make a study of the flight of birds in order to make political prophecies based on the study of individual characters and the circumstances that surround them.[33] Even as he describes divination as both superstitious and erroneous in *On Divination*, Cicero nevertheless acknowledges the importance of the "science" of interpreting signs in order to uphold "Religion and the State."[34] Divination may have been bad metaphysics, according to Cicero, but it was good politics. It is thus understandable that Cicero's account of divination elevates outcomes over causes (*eventa, non causae*). As Quintus observes, "I do not know *why* any of these things happen, but I do grasp *what* happens."[35] The point of divination for Quintus and his fellow Stoics, in other words, is not to explain causes about why something happens, but rather to "give notice of what will happen unless you provide against it" (158). Divination in Cicero, as for many later authors, was thus far more than anti-empirical superstition. It was a way to understand the divine presence or *logos* through consideration of the order of nature in and through which the gods could be made visible. Christian theologians from Augustine to Calvin disparaged such divination but also recognized its contribution to their efforts to read and appreciate the order of nature.[36]

Monstrous Signs

Characters often invoke the gods in both *Julius Caesar* and *Antony and Cleopatra*, though the gods themselves do not appear in either play. In 1.1, for example, Marullus orders the plebeians to disperse, return home, fall upon their knees, and "Pray to the gods to intermit the plague" that is sure to meet the "ingratitude" they have shown to Pompey by celebrating the triumph of Caesar in the streets of Rome (1.1.53–4). But the tribunes are in fact trying to repress the public worship of Caesar during the festival of Lupercal, including their idolatrous worship of Caesar's statues. In the next scene, Casca claims that Marullus and Flavius are "put to silence" for pulling "scarves" from Caesar's statues (1.2.279), suggesting a conflict between the will of the gods and the actions of Caesar. The play thus invites audiences to perform acts of interpretative divination themselves, and our efforts to understand the gods and the influence of religion more broadly depend on efforts to interpret the order of nature. In the process, an aura of anxiety emerges around all signs in the play, even from its opening lines.[37] In questioning the plebeians marching in the streets, the tribune Flavius warns those who follow "mechanical" trades that they must have the "sign" of their professions about them (1.1.3–4). Once he has dispersed the crowd, moreover, Flavius orders his fellow tribune Murellus to disrobe Caesar's statues of "images" that threaten to elevate Caesar "above the view of men" (lns. 67, 74). Efforts to police the social body of republican Rome shift in the second scene of the play to the efforts of Cassius to read the body of Brutus and of both to interpret Caesar as a ruler. Cassius observes that Brutus does not proffer the "show of love" he used to (1.2.34) before offering himself as a mirror or "glass" by which Brutus can see himself (ln. 68). Later, Caesar describes Cassius to Antony as a "great observer" who "looks / Quite through the deeds of men" (lns. 202–3), the mark of someone who is a potential concern for leaders such as Caesar. This attention to the simultaneous power and ambiguity of signs accords with the play's focus on divination as a system of interpretation.

Caesar's overly hasty rejection of the Soothsayer's warnings about the Ides of March in 1.2 joins other important acts of misprision in the play. Many characters struggle to understand dreams, portents, prodigies, and even the time of day, but Caesar himself has less time for such interpretative acts. In the storm scene, Cicero cautions Casca against such misinterpretation of "portentous things," warning him in the spirit of

Academic skepticism that "men may construe things after their fashion / Clean from the purpose of the things themselves" (1.3.34–5). Cassius nevertheless regards portents as the product of a heaven that has "infused" nature with "spirits" that might protect the Republic from tyranny (ln. 69). The dream of Calphurnia on the eve of Caesar's assassination about a statue of her husband spouting blood into the hands of "many lusty Romans" (2.2.78) also becomes a touchstone for understanding the operation of divination.[38] This vision, which is not in Shakespeare's sources, takes place around the same time that Caesar sends augurs to sacrifice an ox that turns out to lack a heart. Initially refusing to change his course in response to either form of divination, Caesar reluctantly agrees to remain at home instead of journeying to the Capitol, but Decius arrives to offer a re-interpretation of the dream as an image of "reviving blood" that will nourish the Roman people (ln. 88). As scholars note, this act of reinterpretation casts Caesar as a Christ-like figure whose blood produces nourishment for his followers, deriving perhaps from the "Fountain of Life" motif in medieval and early modern Catholic manuscripts that figured the eucharistic body of Jesus as a source of spiritual and material sustenance.[39] At the same time, it is important to note that Caesar, unlike Christ, is not a willing participant in ritualistic sacrifice, nor is he historically situated to understand this layer of Christian imagery. In any case, by describing great men flocking to the statue who will "press / For tinctures, stains, relics, and cognizance" (lns. 88–9), Decius makes Caesar into an object of idolatrous worship.

The divination that both precedes and follows Caesar's assassination thus aligns him with contested modes of knowledge and dramatic representation in the play. Most of what readers and audience members learn about Caesar himself arrives second-hand. As such, it is especially liable to misinterpretation, just as Cassius and Brutus misinterpret the scene in the forum in 1.2 in which Antony offers the crown to Caesar. As we learn later (1.2.228), Caesar refuses the crown three times rather than putting it on his head, as Cassius and Brutus had assumed. Adding to the confusion, Caesar takes contrasting positions on the role of providence—in 2.2 he asks "What can be avoided / Whose end is purposed by the mighty gods?" (2.2.26–7), to which Calphurnia responds with her warning about comets sent by the gods to warn princes of impending doom. Caesar displays a comparable attitude a few lines later when he observes that "a necessary end / Will come when it will come" (lns. 36–7). In his very next breath, however, he asks the augurers what they have learned from the ritual animal sacrifice he has asked to be performed (ln. 37). The sense of crisis brought on by the storm and Calphurnia's dream begins to impinge on his own heroic self-regard, even as he announces that he is more dangerous than "Danger" itself (lns. 44–5). This kind of self-aggrandizing gesture, amplified by his later claim that he is as "constant as the northern star" (3.1.61), confirms the fears of Caesar's fellow patricians who feel threatened by his power.

Here and elsewhere, Shakespeare highlights the disjunction between Caesar's idealized descriptions of himself and his own frail mortality. For instance, the play acknowledges Caesar's partial deafness (1.2.213) and epilepsy (1.2.250–1), while Cassius emphasizes his physical fragility in claiming to have rescued Caesar from drowning in the Tiber (1.2.100–14). Such fragility may help to explain Caesar's anxious surveying of the world around him, not just in his reliance on augury but also in his fear, for example, of the "lean and hungry look" of Cassius (1.2.194). According to Cassius, Caesar has grown "superstitious" in relation to dreams and "ceremonies" that he once rejected (2.1.195–7).[40] We see, for example, that Caesar urges Antony and others to "leave no ceremony out" in observing the Feast of Lupercal (1.2.11), in which priests historically struck women with thongs made from the skins of sacrificed animals to increase their fertility. David Kaula links these pronouncements with Caesar's "pontifical manner," his Christ-like desire to be regarded as Rome's "redeemer," but they are more ambiguous, in part because Shakespeare denies his characters the prophetic powers of insight they might otherwise claim.[41]

Cassius downplays stellar influence in telling Brutus that the "fault, dear Brutus, is not in our / Stars but in ourselves that we are underlings" (1.2.140–41). But Cassius may be criticizing divination only to justify his own acts of political violence. He acknowledges the power of divine signs to produce a "monstrous state," but he argues that only individual agents can affect social change. This position reflected the Ptolemaic motto of many early modern astrologers: *vir sapiens dominabitur astris*—"The wise man shall rule the stars."[42] Divination in the political context of Shakespeare's tragedy invites questions about the power of individuals to shape history. Shakespeare also employs specific schools of classical philosophy as a way to distinguish his characters from one another. Casca, for instance, follows the Stoic position that all natural, preternatural, and supernatural signs are produced by the gods to inform humans of their will. He rejects the argument that such signs might stem from natural causes. The skeptic Cicero likewise cautions against the subjective interpretation of divine signs. Ever a careful reader of his sources, Shakespeare attempts to present his Roman characters in the thought and language of their time, even as he also frames his drama through the lens of early modern theology and natural philosophy.

As Cicero indicates in his works on divination, Roman Stoics took divination seriously as a way of understanding the divine order of nature. In spite of his professed Stoicism, however, Brutus does not attempt to interpret the portents, prophecies, and other signs of divination of the play. His only response to the "exhalations whizzing in the air" is to observe that they provide light to read with at night (2.1.44–5). In this regard, he resembles Antony, who likewise shows little interest in portents or other forms of divination in the play. These central antagonists of the tragedy, despite their differences, nonetheless share a vision of politics in which human action rather than divine providence drives history. Brutus's failure to live up to his own Stoic ideals adds a glimmer of humanity to his otherwise highly rigid character, but his indifference to the supernatural events of the storm scene also shows that he is out of touch with both the political and natural orders of the play. In the opening lines of act 2, for example, Brutus expresses uncertainty about the time of day and the date of the month. He believes it to be the first of March (ln. 40), before hearing from his servant Lucius that "March is wasted fifteen days" (ln. 59). Then, as the conspirators gather in the courtyard of his house, a debate breaks out between Decius, Casca, and Cinna about where the sun will rise ("You shall confess that you are both deceived," snaps Casca to his interlocutors at line 104). At stake in this apparently petty argument about the time of day is the relation of the conspiracy to interpretations of the natural world and the role of divine providence in relation to Caesar's assassination. Brutus and other conspirators misunderstand the operation of providence because they fail to give adequate attention to the natural and supernatural portents that surround them, alienating them from the movement of the heavens and with ideas of fortune more generally.[43]

Speaking to himself earlier in the same evening, Brutus admits that he has no "personal cause" why he must destroy Caesar, only a shadowy sense that Caesar "may change his nature" by becoming a tyrant if he is crowned (2.1.11, 13). Brutus views Caesar only as a tyrant *in potentia*, comparing him to a "serpent's egg" that has a natural tendency to do harm after it has been born and more fully developed (ln. 32). But Brutus makes a prophecy about Caesar's inevitable tyranny that does not rely on omens or reading supernatural signs but only on his own, flawed reasoning. These predictions overlook inconvenient facts that question his certainty about Caesar's tyranny, including the ruler's refusal to accept the thrice-offered crown during the festival of Lupercal.

Brutus also fails to perceive the political effects of the assassination of Caesar, especially in agreeing to allow Antony to speak in public after Caesar's death (3.1.228–44) and in his disastrous decision to go to battle against Antony and Octavius instead of waiting for the enemy armies to march first (4.3.204–13). In both decisions, he overrules Cassius, who demonstrates a more accurate understanding of the potential dangers that lie ahead. Brutus's hasty prophecies and inability to discern realities surrounding him reveal the limitations of logic and human reason he otherwise admires.[44] But Cassius is also limited in his ability to anticipate the future or even read the signs of the "monstrous state" he describes. Only members of the audience have the potential knowledge and historical perspective to piece together the signs and wonders of the play.[45] Unlike Shakespeare's characters, we have the necessary knowledge to understand that the fall of Caesar will give rise to a new age of empire that, in turn, fosters the age of "universal peace" in which Christ was born. This kind of knowledge exposes the limitations of divination within the play, while also linking divination with futures yet unknown.[46]

Divine Futures

Most of the scenes of divination in *Julius Caesar* occur prior to the assassination of Caesar himself, echoing early modern accounts that focus on the portents and prodigies warning of Caesar's death. But the act of killing Caesar ruptures both time and Fortune in the play, as Brutus and the conspirators take charge over the contingencies of chance events. In rejecting the proposal of Cassius that the conspirators also kill Antony, Brutus argues that they should be "sacrificers" rather than "butchers" (2.1.166). After the murder of Caesar, Brutus tries to enact a sacrificial ritual by making all of Caesar's killers dip their hands in his blood (3.1.106–8). Antony disrupts and reverses such efforts as he arrives on the scene and shakes the hands of each conspirator, turning the blood ritual into a blood bath and calling on the goddess of discord Ate to unleash the "dogs of war" in Rome (3.1.275). Then, the contrasting funeral orations by Brutus and Antony frame Caesar's death less as the consequence of divination than as the result of human action. Where the portents, prodigies, and omens of the first three acts create an atmosphere of providential history that is one degree removed from human agency, these scenes return political agency to human speech and action, largely stripped of divination. Neither Brutus nor Antony invokes the gods in his funeral oration, though Antony refers to the Christ-like "drops" of Caesar's blood (3.2.192) and alludes to a passage from Luke 19:40 ("If these should hold their peace, the stones would immediately cry out") in a way that further links Caesar's death with the sacrifice of Christ (lns. 223–4). Brutus confines himself to explanations of the immediate past in murdering Caesar for his ambition, while Antony links the past to the future and turns the corpse of Caesar into an object of prophetic anticipation.

In recalling the assassination of Caesar that he did not witness himself, Antony spins a powerful tale that links individual holes made in Caesar's robe with the violent thrusts of his murderers, enfolding his audience into the story of the fall of the great man: "O what a fall was there, my countrymen! / Then I, and you, and all of us fell down, / Whilst bloody treason flourished over us" (3.2.188–90). For the first time in the scene, the plebeians begin to speak of vengeance against the conspirators (ln. 201). In his highly theatrical gesture of reading the will of Caesar, Antony emphasizes Caesar's generosity in leaving his personal walks and orchards to the "common pleasures" of the people (ln. 248). The will functions as both a rhetorical prop for Antony and as a promised vision of the future that motivates political change.[47] After the commoners vow to "pluck down forms, windows, anything" (ln. 257), Antony relishes the "Mischief" he has unleashed upon the world, which can take whatever "course" it likes (lns. 258–9). Anticipating his alliance with Octavius, Antony relishes the idea that "Fortune is merry" and will give the two of them "anything" (lns. 264–5), since Antony himself has shaped Fortune in a politically advantageous manner. Yet the spirit of vengeance conjured by Antony quickly spreads, first in the spontaneous violence against Cinna the Poet by four plebeians who kill him because he has the same name as Cinna the conspirator and then in the broader civil war that consumes the Roman state. The list of political enemies marked for death that Antony, Lepidus, and Octavius compile in the opening scene of act 4 parodies in an obscene way the will of Caesar that Antony has read in the previous scene. Both acts of inscription attempt to shape the future, yet neither is entirely successful.

Conflict emerges between Lepidus, Octavius, and Antony that ends in deadly conflict. Externally, the demands of the military campaign against Brutus and Cassius restrict their ability to establish rule in Rome. Similar forces strain the relationship between Brutus and Cassius, who argue violently in the tent of Brutus over a charge of bribery of one of Cassius's followers (4.3.1–6). Such internecine conflict contains none of the supernatural and providential language of the opening acts. Antony and Octavius cannot even muster allusions to the gods, while Brutus and Cassius do so only in passing (4.3.41, 46, 156).

In his debate with Antony about whether to march toward Philippi or wait for Antony and Octavius to advance, Brutus offers a new image of the power of Fortune:

> There is a tide in the affairs of men,
> Which, taken at the flood, leads on to fortune;
> Omitted, all the voyage of their life
> Is bound in shallows and in miseries. (4.3.219–22)

Brutus's declaration of faith in the future contradicts his use of the metaphor of a "tide" in the "affairs of men," since that language implies multiple, shifting times in which to seize the moment and suggests that every ebb will be met, sooner or later, with an equal and opposite flow.[48] As it happens, Brutus is as misguided about his own agency as Caesar was about his. Both, in fact, struggle against forces they cannot control or even understand. The fateful decision by Brutus to march on Philippi aligns with two other divinatory events in the play, beginning with Caesar's ghost, whom Brutus refers to as a "monstrous apparition" and who appears just long enough to announce that he will see Brutus again at Philippi (4.3.281). Like other forms of divination in the play, this one is enigmatic, though Brutus musters only the most prosaic response—"Well: then I shall see thee again?" (4.3.288)—showing once more surprising indifference toward divination given his professed adherence to Stoic "philosophy" (4.3.144).

For Plutarch, the appearance of the ghost is a clear sign of the judgment of the gods against the murder of Caesar (796), but in Shakespeare his appearance is open to a broader range of interpretations.[49] Unlike many other ghosts in Shakespeare, including the ghost of Hamlet, Sr., Caesar's spirit speaks few words and reveals no insights about the political state of the world or his own condition in the afterworld. On one hand, of course, Caesar's ghost embodies the "spirit" that Brutus hoped to kill without shedding blood. On the other, the ghost announces himself as the "evil spirit" or demonic other of Brutus himself (4.3.286). In both cases, he reminds Brutus of the fact that the conspirators have released Caesar's spirit by murdering him. We do not witness the promised reappearance of Caesar's ghost at Philippi, but Brutus invokes him in his final lines of the play as he runs himself against his own sword, crying "Caesar now be still; / I kill'd not thee with half so good a will" (5.5.50–1).

Shakespeare also foregrounds divination in staging the deaths of Brutus and Cassius. In the confusion of the battle at Philippi, Cassius confesses that he has changed his position on the capacity of portents to predict the future in response to the appearance of "ravens, crows, and kites" that have replaced the place of two "mighty eagles" that had perched nearby the day before and fed from the hands of his soldiers (5.1.75–90).[50] The eagle, which was affiliated with Zeus, and the raven, with Apollo, were the two most significant birds in Greek ornithomanteia.[51] The substitution of raven for eagle conventionally suggests a lessening of divine support, since the eagle was an omen of military victory.[52] This avian vision sways Cassius, at least "partly," away from his previous Epicurean position against portents (ln. 78), and he suggests to Brutus that the gods remain "friendly" toward their own armies against Antony and Octavius, even if the "affairs of men rests still incertain" (lns. 96, 98). The birds who trouble Cassius and his soldiers are not signs in need of interpretation of the sort that Caesar's augurs might have undertaken. Instead, they indicate that the "tide in the affairs of man" that Brutus thought he was seizing is actually helping the other side.

Both the supernatural ghost of Caesar and the appearance of the birds mobilize different forms of natural divination that raise questions about the role of divine providence while also reaffirming that the tide has shifted in Antony's favor in the play. The unpredictable contingencies of Fortune serve the ends of a history that few characters in the play can anticipate. As with the dream of Calphurnia, the apparition of the birds reveals more to the audience than to Cassius. The audience knows that such signs, like the portents and prodigies on the eve of Caesar's assassination, confirm what history has already demonstrated, namely that the murder of Caesar unleashed uncontrollable vengeance that pushed Rome into civil war and then into the new political era of empire. Octavius, who earned the moniker "Augustus" following the defeat of Antony and Cleopatra in 30 BCE, actively promoted the deification of Julius Caesar following his death. As we have seen, the rise of his empire was made possible by the successful control of narratives about divination and itself became significant in rise of Christianity.

The final acts of *Julius Caesar* offer glimpses of these as-yet-unrealized futures by staging the rise of Octavius at the expense of Antony. A subtle but important shift in the relationship between the two men appears in act five when Octavius contradicts Antony's idea about which direction to lead his army: "I do not cross you," Octavius vows, "but I will do so" (5.1.20). Shakespeare also gives Octavius rather than Antony the final lines of the play in which he claims the "glories of this happy day" (5.5.82). Possible examples of such glories include the construction of an imperial Rome, which Shakespeare's audience would have recognized even if his characters did not. According to legend, Augustus inherited a city of bricks and left an empire of marble.[53] While Octavius focuses in his final speech on the "glories" of battle, Antony celebrates Brutus not as a god but as a "man" whose life was "gentle" and in whom the elements of nature were perfectly mixed (5.5.73–4). Antony's final praise for Brutus—"This was a man!" (5.5.75)—may represent a humanist ideal, as David Daniell suggests, but it also looks forward to the *ecce homo* of Pontius Pilate as a statement about the corporeal presence of Jesus.[54] In place of the famous comet that announced the deification of Caesar in his sources, Shakespeare offers the fallen, self-mutilated body of Brutus as a natural rather than supernatural sign. The corpse of Brutus echoes and re-presents the corpse of Caesar, while also anticipating the crucified body of Christ. Cicero's imperative from the storm scene that men interpret natural and supernatural signs only as "the things themselves" is thus inadequate. Shakespeare's Roman plays reveal the impossibility of such an approach, since it is only through knowledge that cannot yet be understood that the meaning of divinatory signs can be interpreted.

Shakespeare may have found the typological pattern of the fall of Rome making way for the rise of Christianity in any number of sources, including the influential *Seven Books of History Against the Pagans* (c. 417 CE) by Paulus Orosius, written at the request of Augustine. Orosius reads the birth of Christ allegorically as a punishment for the "pride" of the Roman republic, particularly the "unjust slaughter" wrought upon Rome and its enemies by Julius Caesar.[55] The typological function of divination also helps to explain why the comet does not appear in *Julius Caesar*. The Incarnation neutralizes pagan divination and rescues Christians from classical anxiety about the contradictory function of the gods. From the perspective of later Christians, the birth of Christ silences the oracles that anticipated his arrival. This Christian afterlife of the assassination of Caesar and the rise of the Roman empire also helps to explain why few characters produce accurate interpretations of the portents, dreams, and prodigies in the play. The pre-Christian characters of Shakespeare's Roman plays see the world, as it were, through a glass darkly. In this sense, Philippi gains in significance as both the place where republican Rome dies on the sword and where Christianity begins its advance as a central location for Paul's evangelical efforts to spread the new gospel of Christ. Such a Pauline context links Caesar's assassination by Brutus with the betrayal of Christ by Judas and the vengeful spirit of Caesar as a typological figure of New Testament spirituality.[56] It also adds a new layer of meaning to the declaration by Titinius and Cassius that Brutus is the "last of all the Romans" (5.3.99).

Racial Divination and the Ends of Tragedy

In *Antony and Cleopatra*, Shakespeare acknowledges the typological context of the rise of Roman Empire by making Octavius observe that "the time of universal peace is near" (4.6.5). Plutarch had noted in his life of Antony that it was "predestined that the government of all the world should fall into Octavius Caesar's hands" (997), though Shakespeare in his treatment of this mode of divination ironically places the pronouncement by Octavius just before his defeat by Antony in the Battle of Actium. In other words, the claim signals Octavius's own lack of knowledge about the future even as it anticipates a future that he cannot predict. In *Antony and Cleopatra*, Antony also begins to take more of an interest in augury, asking the Soothsayer in act 2 to predict, for instance, whether his or Caesar's fortunes will "rise higher" (2.3.15). On hearing the Soothsayer's prediction that Caesar will win the day, Antony orders him to silence but then admits that he is correct, not because the gods are on Caesar's side but only because he tends to beat the "odds," as in cock fighting where Caesar's cocks "do win the battle" (2.3.37, 35). The "boy" Octavius remains throughout the play the antithesis of the aged, pleasure-seeking Antony in his ascetic detachment and instrumental rationality that makes him such a formidable opponent in war.

Yet divination is less central in *Antony and Cleopatra* than in *Julius Caesar*. Although Octavius does not consult agents of divination in conquering his rivals, his servant Dolabella calls him "too sure an augurer" for his prediction that Cleopatra would take her own life (5.2.33). (He turns out to be correct, of course.) The prognosticating Soothsayer also makes only two brief appearances in the play. Although claiming rather grandly that "in nature's infinite book of secrecy / A little I can read" (1.2.8–9), the Soothsayer confines his powers to making vague romantic predictions for two of Cleopatra's attendants in the first scene he appears. In act 2, the Soothsayer predicts Octavius Caesar's ascension over Antony, much to the latter's chagrin. Like Julius Caesar in relation to the augurers and the dream of Calphurnia, moreover, Antony's rejection of the prophecy presages his fall from power. But the arts of divination in *Antony and Cleopatra* do not take a central role in determining the fate of characters. As Antony's follower Scarus notes during a battle at Alexandra, "the augurs / Say they know not, they cannot tell." Such soothsayers "look grimly / And dare not speak their knowledge" (4.12.4–6). Likewise, Cleopatra does not consult an augur or prophet in her desire to know the physical features of Octavia after learning that Antony has married her. Instead, she relies on the biased account of a messenger who is eager to please her and who describes Octavia in unflattering terms. Such an account speaks to the emphasis on gossip, rumor, and hearsay in *Antony and Cleopatra*.

The vanishing influence of divination in *Antony and Cleopatra* leads characters to fashion different narratives of past, present, and future. Antony in particular attempts to read the order of nature in new ways. After he cedes victory to Octavius, for example, Antony explains to Eros the difficulty of knowing oneself in terms of efforts to interpret the clouds:

> Sometime we see a cloud that's dragonish,
> A vapour sometime like a bear or lion,
> A towered citadel, a pendant rock,
> A forkèd mountain or blue promontory
> With trees upon't that nod unto the world
> And mock our eyes with air. Thou hast seen these signs.
> They are black vesper's pageants. (4.14.2–8)

For Antony, the ever-shifting clouds render the *ordo creationis* illegible. These clouds also suggest his own shifting identity. Even though he claims the name "Antony," like the clouds he "cannot hold" such a "visible shape" (4.14.13–14). Such an analogy removes him from the economy of divination in the play. Unlike characters in *Julius Caesar*, his fall is not the result of misreading signs in the order of nature but, rather, a product of that order itself, stripped of its divine creator (though Antony's reference to "black vesper's pageants" may also allude to a theological tradition of ritual and performance with sacred vestiges).

Antony's reference to "black" pageants also highlights the racialized order of nature in the play. The clouds are "black" in Antony's description because they are darkening in color due to the setting sun—in anticipation of the evening vespers. Yet the word joins other racialized terms in the play used to define both Antony and Cleopatra. Philo captures the racial and gendered dynamics of such darkness, for example, in the opening lines of the play:

> Nay, but this dotage of our general's
> O'erflows the measure. Those his goodly eyes,
> That o'er the files and musters of the war
> Have glowed like plated Mars, now bend, now turn
> The office and devotion of their view
> Upon a tawny front. His captain's heart,
> Which in the scuffles of great fights hath burst
> The buckles on his breast, reneges all temper
> And is become the bellows and the fan
> To cool a gypsy's lust. (1.1.1–10)

As Philo suggests, Antony's romantic entanglement with Cleopatra threatens to emasculate him by bending his eye upon the "tawny front" of Cleopatra. Octavius repeats these claims when he mocks Antony for keeping the "turn of tippling with a slave" (1.4.19). The concern of Octavius and others in the play is not just that Antony will become more feminine in leaving behind Mars for Venus, but also that the experience will "stain" him in a physical way—that he will become darker like the "slave" of Egypt. This racial element of divination links *Antony and Cleopatra* to a range of early modern racial discourses that merged forms of natural philosophy and theology, especially in relation to theories of "degeneration," a biblical concept of moral decay within bodies, minds, and souls. In his *Complaint of Peace (Querela pacis)* (1517), for example, Desiderius Erasmus notes that, even though Nature has created us with "reason" and the "will and mind of God," our inherent lust and sin lead inevitably to our "degeneration" into beasts.[57] Decades later, the English doctor William Clever juxtaposed the healthy body as "one of the high and wonderful works of God" with the unhealthy "degeneration of nature."[58] Aristotle had provided an influential account of degeneration in his *Generation of Animals* where he described "monstrosity" in children who do not resemble their parents and thus violate the normative "type" of nature (767b). One common type of such "monstrosity" was the generation of female rather than male children, a misogynistic view that influenced Galen and others who went on to define women as imperfect males in their medical writings.[59] Christianized natural philosophers increasingly justified this classical narrative of degeneration by recalling biblical stories of degeneration in medieval and early Europe.

In the Old Testament, the idea of degeneration appears first in the story of the Fall of Adam and Eve. The disobedience of the first humans in Eden stems from their act of eating the fruit of the Tree of Knowledge, which leads God to curse them and withdraw His presence (Gen. 3:14–19). Such primordial sin flows through the descendants of the first parents, beginning with Cain, whose children suffer for his murder of Abel. This fratricide of Abel leads God to withdraw his presence from humanity, though the eventual birth of Seth renews hope for spiritual reunion with the divine creator (Gen. 4:25–26). Like Seth, Noah appears in Adam's physical likeness, which allows him to find favor with God. But the God of the Old Testament still curses the human race for its general degeneration, while choosing Noah and his family as the sole human survivors of the Great Flood, an act that returns the planet to the "formless and empty" state prior to God's first creation (Gen. 1:2).

In agreeing to God's covenant following the flood, Noah submits to the command to be fruitful and multiply (9:1) in hope that humankind will once again emerge as God's image (9:6). But degeneration returns to curse Noah's own lineage. Chapter 9 of Genesis tells the "generation" of Noah in his three sons—Shem, Ham (also spelled "Cham"), and Japtheth—born to him and his wife after the great flood. These three sons become progenitors of future generations of men who take control of the restored land of the world. We learn in Genesis 9:20 that, after the Great Flood, Noah becomes a farmer, planting vines for grapes. Drinking wine from these grapes, Noah appears drunk and naked before Ham one night, who informs his older brothers. The older brothers cover their naked father, who awakes to learn that Ham has neglected him (or, according to some interpretations, made fun of him or even sexually abused him) and prophetically curses Ham's son Canaan, as the Geneva bible has it, to be a "servant of servants" (9:25). Some commentators used this story of sin and redemption to explain why some of Ham's descendants became slaves. Apologists for the slave trade, for example, often held that Ham's descendants included the inhabitants of sub-Saharan Africa. In early modern England, degeneration began to feature in studies of skin color and physiology, including those undertaken by John Bulwer and Thomas Browne on behalf of the Royal Society.[60] Such studies linked physical, spiritual, ethical, and national traits together under the rubric of degeneration. These early modern accounts of degeneration acquired a more biological orientation, helping to develop new modes of scientific race and accompanying forms of racism.[61]

Later scenes in *Antony and Cleopatra* expand upon this racial dynamic. When Antony abandons battle to pursue the fleeing ships of Cleopatra in act 4, he curses her as a "foul Egyptian" who has betrayed him (4.12.10). Beyond the figurative sense of moral corruption, the word "foul" could also signal something "dull and dirty in color."[62] This racialized accusation carries over into Cleopatra's response to Antony's death. Injured from wounds caused by his own sword, Antony appears before Cleopatra in the tomb, where she frames the event in cosmological terms: "O sun, / Burn the great sphere thou mov'st in; darkling stand / The varying shore o'th' world" (4.15.10–12). The surface meaning of this apostrophe refers to the darkening of the earth that would occur if the sun burned through the celestial sphere in which it was located. (It is fitting for the historical Cleopatra to invoke a Ptolemaic rather than Copernican model of the universe.) But the implication of "darkling stand" is also that Antony himself is a "darkling" who is stained by his amatory desire for Cleopatra herself, one whose soul or shade is also destined for the darkness of the underworld. Like much about Shakespeare's most ambiguous Roman play, the process of racializing Antony and Cleopatra remains elusive, as do the signs of the *ordo creationis*.[63]

Divination thus marks forms of confusion in both *Julius Caesar* and *Antony and Cleopatra*, including uncertainty about claims of authority in relation to the order of nature. Divination also indicates epistemological ruptures in both. Both tragedies challenge Timothy Reiss's claim, in fact, that "tragedy brings about rationality by showing what can be termed the irrational within that rationality."[64] Tragedy in Shakespeare's Roman plays may attempt to realize the inexpressible through performance, but such performance also relies on an ambiguous system of signs in which divine presence remains both "unfathomable and unknown," in Brahe's formulation. Divination is a useful tool for analysis, but only, as we have seen, up to a point. It highlights the necessary forms of ignorance that accompany insights into the order of nature. In Shakespeare's hands, it also employs the *ordo artificialis* in the form of theatrical mimesis to stage questions about divine agency and the legibility of the *ordo creationis* as designed by an invisible and "unfathomable" God.

Notes

1. Astronomers estimate that only four or five comets per century can be seen with the naked eye during the day. For an historical overview of the Great Comet, see C. Doris Hellman, *The Comet of 1577 and Its Place in the History of Astronomy*. See also Hellman's "Bibliography of Tracts and Treatises on the Comet of 1577." For an account of early efforts to understand comets before the advent of telescopes, see Roberta J.M. Olson, "...And They Saw Stars."
2. Aristotle's view of comets, which was dominant for two thousand years before Brahe, was that they originated in vapors in the earth warmed by the sun that then rose through the spheres until becoming solid. The most detailed discussion of them appears in his *Meteorologica* (342a).
3. See Kocher, *Science and Religion in Elizabethan England*, 177.
4. Brahe, *Opera Omnia*, ed. Dreyer, 4:385–6. English translation by Christianson in "Tycho Brahe's German Treatise on the Comet of 1577," 132.
5. Brahe, in Christianson, trans., 132–3, 137. Brahe was certainly no mystic, but his emphasis on the unknowability of the divine creator of the comet puts him in very good company among his contemporaries, as Kevin Killeen has recently analyzed in *The Unknowable in Early Modern Thought*.
6. Nicholas Copernicus, "Preface" to *De Revolutionibus*, ed. Nobis and Sticker, 4. The original Latin reads "mundi, qui, propter nos, ab optimo et regularissimo omnium opifice." Cited in Kenneth J. Howell, *God's Two Books*, 13.
7. See Struck, *Divination and Human Nature*, 1–36.
8. Brahe, in Christianson, 137. For more on the prophetic aspect of early modern astronomy in Brahe and others, see Håkan Håkansson, "Tycho the Prophet."
9. For more on this context, see Brenda Deen Schildgen, *Divine Providence, A History*, 74–5.
10. The observation by Ficino is cited in Sarah Hutton, "Platonism in some Metaphysical Poets," in *Platonism and the English Imagination*, ed. Baldwin and Hutton, 172.
11. For more on the everyday nature of the magic, see the essays collected in *Everyday Magic in Early Modern Europe*, ed. Kathryn A. Edwards.
12. Luther, *A Very Comfortable and Necessary Sermon in These our Days*, sig. C7v. The status of divination is ambivalent in both the Old and New Testaments. Isaiah mocks the "star-gazers" and "monthly prognosticators" who look to the heavens for true revelation (Isa. 47:12–4), for example, whereas those who receive God's wisdom do so without looking for it (see, e.g., Isa. 1:1, Amos. 1:1). Isaiah (8:19, 19:3), Deuteronomy (18:11),

and Leviticus (19:31) also condemn necromancy, while Ezekiel mocks the King of Babylon for stopping at a crossroads to cast lots, consult idols, and perform haruspicy of the liver (21:21). Other biblical passages seem to support certain modes divination, including the dream interpretations of Jacob (e.g., Gen. 31:10–14) and Daniel (Dan. 1:17, 2:1–49), as well as the prophetic dreams of Joseph in Matthew (1:20, 2:13). Aaron, the brother of Moses, also appears to cast lots in deciding on a sacrificial animal for Yahweh in Leviticus 16:7–10, while Joseph performs what seems to be a kind of hydromancy in Genesis 44:5. In Judges, the prophet Gideon looks for and receives a sign from God in the form of dew that appears overnight on a piece of wool or fleece (6:36–40). The enigmatic objects of Urim and Thummin ("light and darkness") placed near the breastplate of Jewish high priests are often sources of divinatory knowledge, used for example by Samuel in selecting a king (1 Samuel 10:20–22) and by Saul in his war against the Philistines (1 Samuel 14:36–7). Talmudic wisdom suggested that the objects were associated with the patriarchs Abraham, Jacob, and Isaac and shone with divine presence to help them make important decisions.

13. For an account of Melanchthon's views of divination and his relation with Luther's horoscope, see Aby Warburg's section on "Pagan-Antique Prophecy in Words and Images" in *The Renewal of Pagan Antiquity*, 600–611.
14. See, e.g., Louise Hill Curth, *English Almanacs, Astrology, and Popular Medicine*. Also see Mazzio, "Coda," in *Spectacular Science*, ed. Chiari and Popelard, 240.
15. This passage appears in the 1640 translation of Bacon's *De augmentis*, trans. Watts, 147.
16. Ibid., 149.
17. Suetonius, *Lives of the Caesars*, 1:136–7 and 1:146–7.
18. Pliny, *Natural History: Books I and II* (2.93–4), trans. H. Rackham, 236–9. See also Emma Gee, *Ovid, Aratus, and Augustus: Astronomy in Ovid's "Fasti,"* 158–9. Romans had been prepared for Caesar's dramatic catasterism by virtue of the cult of divinity or *Divus Iulius* that Caesar himself developed on the model of Romulus, who was "hailed as a god" after his death according to Livy in his *History of Rome* (1.16), trans. B.O. Foster, 1:58–9. The myth also had antecedents in classical Greece and Egypt. For example, the Greek rulers Alexander and Tigranes were thought to have ascended to heaven after their deaths.
19. Plutarch, *Lives of the Noble Grecians and Romans*, trans. North, 796; 792–3. This edition, which is most likely the one that Shakespeare used, is hereafter cited parenthetically in the text by page number.

20. Ovid, *Metamorphoses* (15.834–42), trans. Golding, ed. Forey, 459. Livy's compendium of Roman prodigies—his *Julius Obsequens*—adds additional portents, including a rainbow that pointed directly to Octavius Caesar (later Augustus) on his return to Rome following the defeat of the conspirators, the toppling of a statue of Cicero by a sudden "tornado," and the overflowing of the banks of the river Po. See his *History of Rome*, trans. Schlesinger, 14:308–11. A translation of the Alsatian encyclopedist Conrad Lycosthenes published by Stephen Batman in London in 1581 as *The Doom Warning All Men to Judgment* includes all of the "tokens and prodigies" mentioned in this paragraph, along with additional portents (110–13).
21. Marjorie Garber explores divination in relation to questions of "character and consciousness" in *Julius Caesar* in *Dream in Shakespeare*, 48. John W. Crawford addresses the role of fate and the gods in controlling destiny in the play, including the debate between Stoicism and Epicureanism, in "The Religious Question in *Julius Caesar*." Mark Rose in "Conjuring Caesar" argues that, even if Antony himself is a disenchanted "manipulator of the plebeians," the play itself, "with its ghost, its soothsayer, its prophetic dreams and supernatural prodigies, is not" (297–8). David Carson makes a brief argument for "The Dramatic Importance of Prodigies in *Julius Caesar*," but focuses mainly on the fact that the conspirators overlook the ominous portents in the sky during the garden scene.
22. Shakespeare provides a more nuanced and complex treatment of divination in his Roman plays than in his earlier history plays, where divination signals characters who are potentially demonic, such as the Welsh rebel Owen Glendower in the two parts of *Henry IV*. On prophecies and prodigies in the history plays, see Howard Dobin, *Merlin's Disciples*, 154–65. Noting a general "distrust of Sibyls and Sibylline lore" in Shakespeare, Jessica L. Malay nevertheless acknowledges key differences in the representation of divination in his history plays, tragedies, and Roman plays in *Prophecy and Sibylline Imagery in the Renaissance*, 96–120. Of course, as Moody Prior notes in *The Drama of Power*, it is possible to depict history both providentially and realistically at the same time (17). On the historical representation by poets and astronomers of a comet in 1402 marking Owen Glendower's revolt, see Mark Williams, *Fiery Shapes*, 131.
23. In a classic account of the play's politics, Allan Bloom argues that Roman gods "had to do only with political success or failure and did not give indications of a new transcendent dimension to life beyond that provided by civil society." See "The Morality of the Pagan Hero," in *Shakespeare's Politics*, 78. More recent accounts of the politics of the play downplay divination and the role of the gods. See, e.g., the Althusserian reading by Adrian Phoon in "A Vision Fair and Fortunate." By contrast, Daniel Juan

Gil employs Agamben's theory of biopower in "'Bare Life'." A counterexample to this trend is Robert D. Moynihan, "Stars, Portents, and Order in *Julius Caesar.*" Moynihan argues that divination reinforces a "conservative interpretation" of the play that the assassination of Caesar violates natural law (26), and that Brutus remains "ignorant of meaningful design" in the universe (31).

24. I disagree with David Daniell's claim that Shakespeare has "withdrawn the gods from Rome." Their presence is vital and consequential even though they do not appear. See Shakespeare, *Julius Caesar*, ed. Daniell, 60.
25. Frederic Nausea, *All Blazing Stars in General*, trans. Fleming, sig. C7v.
26. Ludwig Lavater, *Of Ghosts and Spirits Walking By Night*, trans. Harrison, 8. Hereafter cited parenthetically in the text by page number.
27. Isidore of Seville, *Etymologies* (*Etymologiarum sive originum libri XX*, 11.3.1–4). Trans. and cited by Daston and Park in *Wonders and the Order of Nature*, 50.
28. Foxe, *Acts and Monuments*, 913. See also Lycosthenes, *The Doom Warning all Men to Judgement*; Stephen Batman, *New Arrival of the Three Graces*; John Merbeck, *A Book of Notes and Common Places, with their Expositions.* See also Sheltco à Gevern, *Of the Ende of This World.*
29. Anon. *History of Strange Wonders*, sig. A7.
30. Camerarius, *History*, sig. D2v. On the religious implications of monstrous-birth broadsides, see Jennifer Spinks, *Monstrous Births and Visual Culture in Sixteenth-Century Germany*, 54–5 and 74–6.
31. See Frederick E. Brenk, *In Mist Apparelled*, 186, 210.
32. English translations of these tracts were not printed in English until 1683, but French translations likely circulated throughout England earlier, together with possible manuscript translations. Camerarius cites *On Divination* in his *History of Strange Wonders* in relation to "monsters and wonders" that demand interpretation, while a 1581 translation of a treatise on "strange prodigies" by the Alsatian humanist and encyclopedist Conrad Lycosthenes alludes to the same Ciceronian text with reference to a farmer who suddenly heard the voice of "child…old in wisdom" while plowing his fields. See Camerarius, sig. B2, and Lycosthenes, *The Doom Warning All Men to Judgment*, 46.
33. See Jerzy Linderski, "Cicero and Roman Divination," 12–13, and "The Augural Law."
34. Cicero, *Of Divination* (2.12), in "*The Nature of the Gods*" and "*On Divination*," trans. C.D. Yonge, 211. All English quotations from these two works by Cicero are to this translation, hereafter cited parenthetically in the text by page number. Cicero's original Latin reads "Ut ordiar ab haruspicina, quam ego rei publicae causa communisque religionis colendam censeo." See the Loeb edition (LCL 154), trans. W. A. Falconer, 400.

35. Cicero, *De divination* (1.16). "Cur quidque fiat, ignorem, quid fiat, intellego," trans. Falconer, 240.
36. See, e.g., Schildgen, *Divine Providence*, 54–5. See also Edward Morgan, *The Incarnation of the Word*, 118.
37. The play was likely first performed in 1599 at the newly constructed Globe theater. Using both internal and external evidence, for example, Stephen T. Sohmer argues the play was first performed on the summer solstice of 1599 in *Shakespeare's Mystery Play*. Cf. *Julius Caesar*, ed. Daniell, 2 and Chambers, *The Elizabethan Stage*, 2:364–5.
38. For a psychoanalytical account of the relation between the interpretation of dreams and the construction of character, see Marjorie Garber, "Dream and Interpretation: *Julius Caesar*," in *William Shakespeare's "Julius Caesar,"* ed. Bloom, 43–52.
39. See Jack Heller, "Your Statue Spouting Blood."
40. The historical Caesar seems to have been quite willing, on numerous occasions, to reject or reinterpret unfavorable divination, as for instance regarding signs related to his departure against Pompey and in his campaign in Africa. See, e.g., Stefan Weinstock, *Divus Julius*, 98, 116, and 342.
41. David Kaula, "Let Us Be Sacrificers," 212. As these words suggest, Kaula also locates a central "typological mode of thought" in Shakespeare's play, but his focus on Caesar as an emblem of Christ overlooks larger typological structures that depend on divination.
42. The phrase was attributed to Ptolemy, although it does not appear in his extant writings. Variations do occur, however, in the poetry of Dante and Gower, as well as in the *Summa Theologica* of Thomas Aquinas and in many works of astrology throughout the thirteenth and fourteenth centuries. See Theodore Otto Wedel, *The Mediæval Attitude Toward Astrology*, 68, 135–7.
43. It is possible that the conspirators are seeing the portents in the sky without knowing it, as Carson argues in "The Dramatic Importance of Prodigies in *Julius Caesar*," 179. Brutus returns to the scene before Decius, Cinna, and Casca resolve their debate over the location of the sunrise.
44. See Ernest Schanzer, "The Tragedy of Shakespeare's Brutus."
45. Julian Rice argues that the request by Brutus to his fellow conspirators that they judge by their senses is ironic in a play in which such judgment is so often misguided. Rice suggests that the illusions of Brutus and others demonstrate "the play's general Pyrrhonism, which finds man's knowledge and beliefs to be determined more by fear and hope than by reason." See "*Julius Caesar* and the Judgment of the Senses," 247.
46. Maurice Hunt argues that the play aligns Caesar and Christ in a "positive" way that is also occasionally "parodic." He also claims that the opening

scene, in its allusions to cobblers and its pun on soles/souls, establishes important "Christian overtones" in the tragedy. His main argument is that Shakespeare develops for the first time in *Julius Caesar* a Christian sense of the soul as empathetic to the "frailties and sufferings of others." See "Cobbling Souls in Shakespeare's *Julius Caesar*," in *Shakespeare's Christianity*, ed. Batson, 112, 126.

47. In Plutarch, it is unclear if the will is read initially in the senate and then by the commoners or if it is discovered by the commoners. In either case, the reading of Caesar's "testament" is what ignites the violence of the plebeians toward the conspirators and causes them to burn the body of Caesar before lighting the houses of the conspirators on fire with "firebrands" from the funeral pyre (795).
48. Such a "tide" of time is part of a larger pattern of rising and falling action that John Velz describes in his essay on the "Undular Structure in *Julius Caesar*."
49. John Roe suggests that Plutarch both attributes the punishment of the conspirators to the gods and designates Caesar's ambition as the cause of his downfall in "'Character' in Plutarch and Shakespeare," in *Shakespeare and the Classics*, 176.
50. As Allan Bloom remarks, "the Cassius of acts IV and V is a very different man from the one we saw previously. He has become superstitious. He now believes that there are spirits and a providence in the world." But Bloom ignores the larger function of divination as it influences characters, politics, and the divine order of nature in the play. See *Shakespeare's Politics*, 103.
51. See Dillon, *Omens and Oracles*, 3.
52. Ibid., 147.
53. See, e.g., Kathleen S. Lamp, *A City of Marble*.
54. *Julius Caesar*, ed. Daniell, 321.
55. Orosius, *The Seven Books of History Against the Pagans,* trans. Deferrari, 266–7, 288.
56. Roy Battenhouse briefly outlines a "dark-shadowing of Christian pattern" in the play, reading the Feast of Lupercal as a parody of the Christian Passover and the assassination of Caesar in terms of the betrayal of Christ in *Shakespearean Tragedy*, 92–3. This shadowing receives more substance in Nicholas W. Knight, who sees the play primarily as a form of Christian allegory in "*Julius Caesar*: A Case of Pre- and Post-Christian Story." My reading stresses the hybridity of pagan and Christian meaning and the ambiguities of divination, resisting such overt allegory.
57. Desiderius Erasmus, *The Complaint of Peace,* trans. Paynell, sigs. A5v, A7.
58. William Clever, *The Flower of Physic*, 116.

59. Rosemarie Garland Thompson notes that "Aristotle initiates the discursive practice of marking what is deemed aberrant while concealing what is privileged behind an assertion of normalcy." See *Extraordinary Bodies*, 20. For more on this topic, see Fiona Mitchell, *Monsters in Greek Literature*, 155–90, esp. 181–4.
60. On the science of skin color in Bulwer and seventeenth-century Baconian science, see Cristina Malcolmson, *Studies of Skin Color in the Early Royal Society*.
61. See, e.g., Anu Korhonen, "Washing the Ethiope White," in *Black Africans in Renaissance Europe*, ed. Earle, 94–112.
62. *Oxford English Dictionary*, 2d. ed., s.v. "foul," adj., def. 5.b.
63. Janet Adelman notes, for example, that "almost every major action in the play is in some degree inexplicable" in *The Common Liar*, 15. Ronald R. MacDonald makes a similar point when he says that "the meaning of virtually every tear in the play is ultimately undecidable" in "Playing Till Doomsday," 81.
64. Timothy Reiss, *Tragedy and Truth*, 284. For a useful critique of this position in relation to other theories of tragedy, both classical and modern, see Terry Eagleton, *Sweet Violence*, 18–21.

CHAPTER 8

King Lear and the Orders of Nature

Like *Julius Caesar* and *Antony and Cleopatra*, *King Lear* combines classical, medieval, and early modern modes of natural philosophy and theology to frame nature as an object of inquiry and source of theatrical mimesis. The play includes more allusions to "nature" and its cognates than any other by Shakespeare.[1] The definitions of this crucial term proliferate as the drama unfolds, encompassing questions about filial duty, political power, sexual reproduction, the power of language, and the *ordo creationis*. In revising the older *Leir* story that ended with a joyful reunion between Lear and Cordelia, Shakespeare complicates simplistic dualisms between nature and the supernatural, between knowledge and disbelief. The shocking death of Cordelia in Shakespeare's version heightens the pathos of the tragedy's final moments and provides a foil of sorts for Lear's own tragic death. As Gloucester notes in the Folio version of the play, moreover, the central plot of Lear's dramatic fall from divine monarch to homeless wanderer enacts a fall "from the bias of nature" (F 1.2.101–2), where "bias" means balance and draws on the sport of lawn bowling. Such a fall brings Lear and his companions face to face with the state of nature in its barest form, while also generating forms of redemptive materialism.

A. Kitch, *Shakespeare's Theater of Nature*, Palgrave Studies in Literature, Science and Medicine,
https://doi.org/10.1007/978-3-031-78082-0_8

Christian Materialism

The uncertain role of religion in the play has been a persistent problem for readers and viewers over the centuries since its original composition (and likely revisions by Shakespeare in his final years). Set in a mythical, pre-Christian "Britain," *Lear* includes both pagan deities and Christian allusions. In his pioneering analysis of the play in *Shakespearean Tragedy* (1904), A.C. Bradley observes that the pagan gods who populate the drama inspire "lust, pride, hardness of heart, the insolence of office, cruelty, scorn, hypocrisy, contention, war, murder, [and] self-destruction" in central characters, yet the suffering they inflict also creates the possibility of spiritual redemption in the second half of the tragedy. Bradley posits a divide between "good" and "bad" characters, claiming that such "good" characters "are seen growing better through suffering, and the bad worse through success."[2] Yet he also praises Shakespeare for refusing to make Christian beliefs a "constant presence," arguing that Christian readers must hold their beliefs in "temporary suspension" when "immersed in a Shakespearean tragedy."[3]

Subsequent accounts of *Lear* have both expanded upon and challenged Bradley's narrative of Christian redemption that identifies Cordelia as a Christ figure and locates moral lessons about patience, grace, and providence in the play. G.K. Hunter, for instance, identifies a central "process of restoration by which humanity can recover from degradation" in the Lear plot.[4] Likewise, G. Wilson Knight links Lear with the biblical Job, while Robert Speaight describes a "redemptive process" in the tragedy.[5] More recently, A.D. Nuttall has located forms of "Christian love" in the "dreadful pre-Christian world" of the play.[6] But other critics have challenged claims of spiritual redemption, describing the play in starker terms of existential negation. Writing in the immediate aftermath of World War II, for example, Jan Kott regards the play as a parody of Job by puppets ignorant of their own buffoonish identities. In *Shakespeare Our Contemporary* (1964), Kott describes *Lear* in these terms:

> In Shakespeare's play there is neither Christian heaven, nor the heaven predicted and believed in by humanists. *King Lear* makes a tragic mockery of all eschatologies: of the heaven promised on earth, and the heaven promised after death; in fact—of both Christian and secular theodicies; of cosmogony and of the rational view of history; of the gods and natural goodness, of man made in the 'image and likeness'. In *King Lear*, both the medieval and the renaissance orders of established values disintegrate. All that remains at the end of this gigantic pantomime is the earth—empty and bleeding.[7]

Kott's vision of an "empty and bleeding" earth at the center of the drama reflects the tone of many post-war productions of the play and has attracted numerous followers.[8] Jonathan Dollimore, for instance, extends Kott's reading by claiming that *Lear* refuses redemption by making "visible social process and its forms of ideological misrecognition."[9] Richard Halpern goes even further in his argument that "*Lear* is perhaps the most demystifying, indeed materialist, of Shakespeare's meditations on kingship, which the play refuses to hedge about with even a hint of divinity."[10] But these cultural materialist readings rely on an ahistorical opposition between Christianity and materialism that ignores both the ways in which matter itself can be divine and important debates of classical, medieval, and early modern natural philosophy and theology that inform the play. As this chapter argues, the barren landscape of the tragedy's second half teems with modes of divinity and shifting forms of representation that reframe the order of nature at the core of Shakespeare's vision.[11]

More recently, several scholars have emphasized *Lear's* links to early modern science, focusing in particular on Shakespeare's direct allusions to astrology, mathematics, anatomy, and meteorology. In a probing analysis, for example, Mary Thomas Crane suggests that the play foregrounds the "human experience of materiality" and reveals the limits of historical efforts to "explain the invisible and counter-intuitive workings of matter."[12] Identifying central questions about weight, smell, and divisibility in the tragedy, Crane mobilizes theories of matter as debated by a range of early modern natural philosophers, especially Thomas Harriot. Extending Crane's account, I situate *King Lear* within different strands of natural philosophy while highlighting important religious contexts and problems of theatrical representation that she overlooks. I suggest that repeated allusions to modes of natural philosophy in the tragedy mirror shifting modes of dramatic representation, creating a larger pattern that moves from transcendental (pagan) metaphysics positing a unity between language, nature, and divine creation to a model of immanent Christian redemption centered on the "natural" body and language of Lear, Edgar, and Cordelia. Such a reading challenges a conventional opposition between religious and materialist readings of the play that has defined critical accounts since Bradley, joining Michael Witmore in suggesting that Shakespeare favors "a view of the world in which order and change are seen to emerge holistically from things themselves (immanence) rather than being localized in certain metaphysically isolated pockets of the universe (punctualism)."[13]

William R. Elton's still-influential study of *"King Lear" and the Gods* (1966) informs my reading. Elton identifies two main attitudes toward divine providence in Shakespeare's lifetime: (1) Epicurean atomism in which God governs the operation of the universe but "in ways bafflingly inscrutable and hidden to human reason" and (2) a Calvinist assertion of *deus absconditus* in which God is absent from creation all together, completely "incomprehensible and unappealable."[14] Locating both positions in a wide range of English and French sources, Elton reads the play as a complete rejection of Christian providence that captures such skepticism by challenging doctrines of faith, including the immortality of the soul and the elevation of humans over animals. He finds evidence for such skepticism in the play when Lear, for example, questions rather than denounces God's providence at Cordelia's death and fails to consider the spiritual possibilities of an afterlife (54). Despite the moments of Christian faith that surface in the play, Elton ultimately reads Shakespeare's work as "a syncretically pagan tragedy" that calls into question a range of Christian tenets (338).

Although Elton regards the play as a genuine effort to represent pagan theology, he also sees Shakespeare as necessarily "conditioned" by the Christian beliefs of his own period. He defines four central strands of religious belief, each embodied by a different character or set of characters: (1) Edgar and Cordelia represent a *prisca theologia* or original faith that foreshadows Christianity, (2) the villainous characters of Regan, Goneril, and Edmond (Edmund in the Quarto) personate forms of atheism, (3) Gloucester stands for a "superstitious pagan," and (4) Lear moves from "pagan belief to disbelief" (336). Elton gives special attention to Edmond, who claims "Nature" as his own "goddess." Elton groups Edmond with the other atheists of the play in suggesting that Edmond privileges the law of nature (*nomos*) over the idea of nature as a generative principle (*physis*). Elton also situates the play more broadly within a historic shift from nature to convention and as part of the larger "rediscovery of 'nature' itself" in the period that challenged previous faith in the "universality of God's law" (126).

Elton follows a Burckhardtian emphasis on Renaissance individualism that aligns Edmond with both the birth of natural law in early modern England and a nascent strand of libertinism that he also locates in narratives of Don Juan in the seventeenth century (138–46). Yet his invocation of *physis* in the context of a conflict between traditional Christianity and the revival of Epicurean atomism overlooks other important strands of classical, medieval, and early modern natural philosophy that influences the play. I agree with Elton that Shakespeare mediates the play's complex and invisible powers of divinity through ideas of nature and I find his distinction between nature as convention or *nomos* and nature as material creation or *physis* productive. But my own interpretation expands on and complicates his account of the theology in the tragedy by identifying additional and different meanings of nature, some of which are in tension with one another. As I see it, "nature" has at least six overlapping definitions in *King Lear*: (1) as a norm governing relations within a family, as in Lear's relation to his daughters and Edmond's relation to his father (e.g., when he claims to Gloucester that Edgar has an "unnatural purpose" of attempted parricide [Q 2.1.50 / F 2.1.52] or when Lear suggests that Regan knows better the "offices" or duties of "Nature" in observing the proper filial respect toward him [Q 2.2.344 / F 2.2.357]); (2) as a form of inherent identity, as when Edmond acknowledges that his brother's "nature" is "so far from doing harms" that he suspects others of equal goodness (Q 1.2.154 / F 1.2.158) or, near his death, when he feels inclined to do "some good" "despite" his "own nature" (Q 5.3.239–40 / F 5.3.218–9); (3) as a divine force (usually capitalized) that governs the universe, including the "goddess" whose "law" Edmond claims fealty (Q 1.2.1 / F 1.2.1), to which Gloucester attributes the eclipses prophesying mutiny and discord in families, courts, and countries (Q 1.2.95–100 / F 1.2.94–99), and whom Lear invokes in cursing Goneril with sterility (Q 1.4.258–62 / F 1.4.241–4); (4) as a generative force of biological life that attributes human life to the *ordo creationis*, as when Kent insults the Steward of Cornwall by claiming that "Nature disclaims" its part in making him because of his cowardice (Q / F 2.2.49) and when the blinded Gloucester calls the mad Lear a "ruined piece of nature" (Q 4.6.131 / F 4.5.132); (5) as a measure of the elemental life force of a human being, as in Regan's warning to Lear that "Nature in you stands on the very verge of his confine" (meaning his life is about to expire, Q 2.2.312 / F 2.2.326) or when Lear believes that only daughters could be capable of subduing "nature" to such a "lowness" as that he observes in Edgar when disguised as Poor Tom (Q 3.4.62 / F 3.4.67); and (6) as the condition of fallen mankind in a specifically Christian sense, as when a servant to Cordelia calls her a "daughter / Who redeems nature from the general curse" (Q 4.6.192–3 / F 4.5.197–8).

As these definitions suggest, many instances of nature in *Lear* mix religious, scientific, and material modes of mimesis, drawing on hybrid forms of paganism, Christianity, and natural philosophy. A seventh meaning of nature encompasses the familiar dyad of art and nature, as when Lear in his madness refers to his monarchical power of "coining" in the Quarto and argues that "Nature's above art in that respect" (F 4.5.85–88). Here and elsewhere, a vision of nature as an artificial construction emerges in the play as part of its attempt to stage the order of nature in and through theatrical representation. Lear's fall from power is both materially and spiritually transformative. In wandering the fields of England, for example, both Lear and Gloucester are marginalized and reduced subjects, instances of Giorgio Agamben's *homo sacer* who are "excluded from the religious community and from all political life."[15] But both also interact directly with the spiritually transformative order of nature that is the subject of shifting modes of natural philosophy, including meteorology, anatomy, the arts of perspective, and medicine. This state of nature marks an important shift in the play from pagan to Christian ethics, epistemology, and metaphysics while also providing the raw material for experiments with modes of performance. For example, Gloucester with the help of his disguised son Edgar overcomes his suicidal despair by means of a stage device that employs a trick of visual perspective, while Lear in his madness both explores new genres of language and makes himself open to natural remedies of plants that Cordelia and the doctor use to repair the "breach" of his "abused nature" (Q 4.7.16 / F 4.6.14). Through such acts of herbal healing, Lear achieves a mode of spiritual "restoration" by taking "medicine" that provides some "repair" for the "violent harms" he has experienced in the play (Q 4.7.28–30 / F 4.6.24–6).

Forms of Nothing

As the play opens, Lear has already made his decision to divide the kingdom of Britain into three parts, but his act of deposition requires a performative ritual in which his three daughters must participate:

> Which of you shall we say doth love us most,
> That we our largest bounty may extend
> Where nature doth with merit challenge? (F 1.1.49–51)[16]

Lear's verbal experiment not only tests his daughters—it also establishes (at least in the Folio text) a central relationship between "nature" and "merit," by which he means the performative power of language to express love. Lear assumes that his daughters will speak naturally and from the heart in a way that will affirm their worthiness to inherit the land that has already been set aside for them. He thus binds filial obligation to natural law as represented by spoken words. Regan and Goneril obey their father's will by competing with one another to proclaim their love for him, but Cordelia refuses to go along with Lear's patriarchal imperative, claiming that she cannot "heave" her "heart" into her "mouth" (Q lns. 77–78 / F lns. 89–90). Questioned by her father as to what she might say that is even "more opulent" than what her sisters have professed, Cordelia responds with the one-word answer: "nothing" (Q lns. 73, 75 / F lns. 84–5).

Lear's angry retort—"Nothing will come of nothing"—shifts the tone of the play while also invoking metaphysical claims from classical natural philosophy (Q 1.1.76 / F 1.1.88).[17] Most editors interpret Lear's reply as a version of the proverbial phrase *ex nihilo nihil fit* that outlines the basic necessity of God's intervention in order to create the universe out of nothingness. Yet no editor or scholar I know of has attributed the phrase to its specific source in the pre-Socratic philosopher Parmenides.[18] Best known in the history of philosophy for the dialogue of Plato that bears his name, Parmenides was a mentor to Zeno and a major figure in ancient Greek thought who elevated metaphysics to new levels.[19] Indeed, some philosophers claim that he inaugurated the tradition of Western metaphysics itself by seeking to banish non-Being from Being and to base all philosophical inquiry on a concept of *logos*. Unfortunately, only fragments of a single work by Parmenides survive as part of a long poem entitled "On Nature" that includes revelations from a goddess named "Truth" about the nature of the cosmos. The aphorism *ex nihilo nihil fit* emerges from that text as a central part of its monistic account of the cosmos that denies the existence of non-being and binds thought and language together.[20] The phrase helps to define the metaphysical plenitude and constancy of the *ordo creationis*, describing an uncreated and indivisible universe in which speaking, thinking, and nature are unified.[21]

In quoting *ex nihilo nihil fit*, then, Lear expresses his belief in the unity of nature, being, time, and language (*logos*). For Parmenides, this *logos* is the transcendent product of divine revelation that also invites forms of scientific inquiry.[22] Plato's portrait of Parmenides in the dialogue by that name mainly addresses logical problems related to the theory of forms. But Aristotle challenged Parmenides in developing his account of an "unmoved mover" who initiates change in all created matter.[23] In his account of that matter, Aristotle posits a time and space of non-being overcome by an initial act of creation. Later Christian theologians adapted his principle of creation to focus on the modes of material change in the universe, reconstructing God's relationship to the *ordo creationis* in new ways. Lear's rejection of Cordelia's nothingness thus affirms the plenitude of nature, invoking the nothingness of the Parmenidean *ex nihilo nihil fit* only to disavow it. This nothingness brought on by Cordelia's refusal to play her father's language game ultimately leads to the unravelling of Lear's power and, eventually, his mental and physical being. Such unraveling reproduces the historical process by which Aristotle and later Christian theologians developed new models of the cosmos that posit God as an unmoved mover. Over the course of the tragedy, however, Lear comes to understand nature, time, and language from new perspectives, embracing the kind of change that Aristotle describes, for example, in his treatment of generation and that both Epicurus and Lucretius make central to their accounts of an atomistic universe.

In response to Cordelia's "nothing," Lear disowns her and divides her portion of land between her two sisters. Even as Kent steps forward in a failed effort to stop him, Lear retains his faith in the world-shaping and divine power of his own language. Invoking the goddess Hecate, Lear justifies his condemnation of Cordelia in relation to the "sacred radiance of the sun" and "all the operation of the orbs" or planets that were thought to produce physical and spiritual life in the universe (Q 1.1.94–7 / F 1.1.106–9). Lear's theologically oriented curses of Cordelia, Kent, and Goneril likewise reveal his unshaken faith that language, political power, and the order of nature are united together in one seamless entity with himself. But challenges to Lear's vision of the universe continue to mount in the play, together with new questions of natural philosophy that accompany shifts in theatrical representation.[24]

Lear continues to swear by the gods in his rejection of Kent's request for leniency toward Cordelia:

> That [Since] thou hast sought to make us break our vows,
> Which we durst never yet; and with strained pride
> To come betwixt our sentences and our power,
> Which nor our nature nor our place can bear,
> Our potency made good, take thy reward. (Q 1.1.155–9 / F 1.1.166–70)

Here it is Apollo rather than Hecate who underwrites Lear's faith in the unity of words, gods, and power, not to mention the natural bond between his biological and social being. But the King of France points out that Lear's rejection of Cordelia is "unnatural" (Q 1.1.206 / F 1.1.217), adding to the growing skepticism about Lear's logocentric faith in cosmic unity. This metaphysical ideal strains even further when Goneril and Regan confront their father about his need for retainers. His daughters' rejection of his demands for respect and kingly privileges comes to a head in Cornwall's decision to place Kent (disguised as Lear's servant Caius) in the stocks, inducing madness. Lear demands an explanation from Cornwall and Regan as to why his servant has been punished. Learning from Gloucester that Cornwall has a "fiery quality" that is "fixed / In his own course," Lear explodes:

> The King would speak with Cornwall. The dear father
> Would with his daughter speak, commands, tends service.
> Are they informed of this? My breath and blood!
> [...] Tell the hot duke that—
> No, but not yet; maybe he is not well.
> Infirmity doth still neglect all office
> Whereto our health is bound. We are not ourselves
> When nature, being oppressed, commands the mind
> To suffer with the body. (Q 2.2.265–6, 273–80 / F 2.2.272–3, 280–88)

A different sense of "nature" emerges in Lear's imaginary scenario about why Cornwall will not speak with him—a nature that links the sick body to the mind. Such a nature accompanies Lear's sudden loss of authority and leads him to question his previous faith in the unity of his language with a divine order of nature. Indeed, Lear regards Cornwall's refusal as a threat to his own identity. In response, he attempts to reassert the unity of speech, physical generation, service, mind, body, and nature. Like the God of Genesis, he even indicates that he has shaped his progeny out of his own breath. But his failure to command others is also a failure of his own body within the order of nature.

As Regan remarks when she arrives a few lines later, Lear's outburst also reveals failures of nature within himself. As she puts it, "Nature" in her father "stands on the very verge / Of his confine" (Q 2.2.312–13 / F 2.2.326–7). Regan reminds Lear of the form of nature that has created him and continues to prop him up materially, urging him to consider himself as connected to rather than dissociated from the *ordo creationis*. But Lear does not receive this message, nor does he achieve introspection at this moment. In acknowledging his own fleshly condition, moreover, he lashes out at the source of such uncomfortable knowledge—his own flesh and blood:

> [Now] I prithee, daughter, do not make me mad.
> I will not trouble thee, my child; farewell.
> We'll no more meet, no more see one another.
> But yet thou art my flesh, my blood, my daughter,
> Or rather a disease that lies within my flesh,
> Which I must needs call mine. Thou art a boil,
> A plague sore, an embossed carbuncle in my
> Corrupted blood. But I'll not chide thee.
> Let shame come when it will; I do not call it.
> I do not bid the thunder-bearer shoot. (Q 2.2.385–94 / F 2.2.398–407)

Lear disavows the older model of divine nature in which the gods who govern the universe and mete out punishment stand ready to be commanded, perhaps sensing that his flesh—his spiritually "corrupted blood"—is becoming the ground for his future, fallen self. As he acknowledges in the midst of the "all-shaking thunder" at the beginning of the storm scene, "Nature's molds" are physically frail and subject to "crack[ing]" (Q/F 3.2.6, 8). This thunderous storm in act 3 marks a dividing line in the play—from that point forward, Lear's mind and body encounter new forms of nature that do not accommodate his earlier theories about them.

Lear also projects his own anxieties of fleshly frailty onto his daughter with characteristic misogyny, perhaps recognizing that his own body has become more closely tied to material nature than he had previously thought. He fears that he has been physically invaded by the "mother" (i.e., hysteria) that "swells up" toward his heart as a form of "*Hysterica passio*" (Q 2.2.233–4 / F 2.2.236–7). Lear's fall thus reverses a conventionally Neoplatonic model of ascension from body to spirit as he moves from monarch to "child-changed father" whose "abusèd nature" must be cured by the administrations of his daughter (Q 4.7.16, 18 / F 4.6.14, 16). Adopting familiar early modern assumptions about the link between materiality and femaleness (with a pun on *mater* and matter), he passes into the order of nature unwillingly. The unity he had earlier presumed between sign and signified, between word and divine justice, begins to dissolve, replaying the historical movement from paganism to Christianity, even as, ontogenetically, Lear crawls backward toward infancy.

The apparently barren landscape in which Lear dwells for the remainder of the play both signifies and enables this transition, which in theological terms echoes the self-emptying or *kenosis* in Christ's passage from God to man (Philippians 2:7). Bradley aptly describes the emptiness of this space as giving "the feeling of vastness, the feeling not of a scene or a particular space but of a world."[25] This non-particular world becomes the ground for forms of discourse that have not inhabited earlier scenes at court and at the royal estates of Lear's daughter. It is a space for testing the virtues and horrors of the *ordo creationis* in relation to new forms of mimesis, a space in which Lear ultimately proves Kent wrong in his argument that "Man's nature cannot carry / Th'affliction nor the fear" (Q/F 3.2.48–9), but Lear too must learn how to disown his misperceptions about nature. His claim upon encountering Poor Tom that "nothing could have subdued nature / to such a lowness but his unkind daughters" recognizes the process of natural subduction that is occurring to him but misrecognizes its source (Q 3.4.62–3 / F 3.4.67–8). Lear comes to know himself, if only slightly, in and through his immersion in a newly material nature that grounds both his madness and forms of self-knowledge. It is also the site for new modes of natural philosophy as Lear enacts his own return to nature.

Lear's fall also introduces a division in the play between characters who believe that nature and convention are united and those who do not.[26] His entry into the void indicates that his status and power are no longer guaranteed by the Parmenidean order of nature of the opening scenes.[27] The main subplot of the play featuring the bastard Edmond's efforts to seize the land and inheritance of Gloucester from his "legitimate" brother Edgar shapes this evolving recognition. Edmond's conversation with Gloucester and Kent in the opening lines of the play occurs in the shadows of Lear's court, as Gloucester draws explicit attention to Edmond's illegitimacy and contrasts him with Edgar, conceived "by order of law" (Q/F 1.1.18). Gloucester expresses his own joy in his transgressive act of copulation, inviting Edmond soon after to shun conventional law and express his fealty to the "goddess" of Nature:

> Thou, Nature, art my goddess; to thy law
> My services are bound. Wherefore should I
> Stand in the plague of custom and permit
> The curiosity of nations to deprive me,
> For that I am some twelve or fourteen moonshines
> Lag of a brother? Why "bastard"? Wherefore "base,"
> When my dimensions are as well compact,
> My mind as generous, and my shape as true
> As honest madam's issue? Why brand they us
> With "base"? With "baseness," "bastardy"? Base? Base?
> Who, in the lusty stealth of nature, take

More composition and fierce quality
Than doth within a dull, stale, tired bed
Go to th' creating of a whole tribe of fops
God 'tween a sleep and wake? (F 1.2.1–15)

Like Cordelia, Edmond rejects custom, but where Cordelia contradicts her father in a very public way and receives a very public punishment for doing so, Edmond vows his services to the goddess "Nature" in private, setting in motion a series of deceitful tricks to convince Gloucester that Edgar is trying to kill him (Q / F 1.2.16). Edmond's invocation of this goddess differs from the transcendent metaphysics of Lear, framed as it is by his own fleshly desires. Edmond grounds his concept of natural law not in the transcendent order of nature but, rather, in his own sexual drive, which he calls the "lusty stealth of nature." His obsession with natural desire stems from his status as a bastard, which Gloucester invokes in the opening lines of the play as a problem of inheritance and status, but which Lear later reframes in act 4 when he impersonates a magistrate in a fit of madness and pardons a man for adultery, arguing "Let copulation thrive" (Q 4.6.113 / F 4.5.112). Edmond justifies his violation of the protocols of social order—that, as Leonard Wright notes in *A Display of Duty* (1589), "Youth by law of Nature are bound to honor, reverence, and obey their ancients"—in a Machiavellian plot that invites us to consider how Lear's own seemingly devout adherence to law and custom might also conceal at its heart a will to power and misappropriation of nature.[28]

Edmond's embrace of nature as *nomos* also marks the divide between nature and justice in the tragedy. For Lear in the early scenes, such *nomos* takes the form of the "goddess" of Nature whom he invokes when cursing Goneril with sterility and, shortly thereafter, hopes that she will give birth to a "child of spleen" as a form of "disnatured torment" (F 1.4.248–9, "dis-utered torment" in Q 1.4.266).[29] Lear's misogynistic outburst stems, in part, from his increasingly misguided faith in his own actions as natural.[30] He understands his power as monarch in terms of his capacity to call on the gods and the nature that they control to support his positions in relation to Cordelia and Kent. Lear's more general movement in the play from divinely supported monarch to "bare, forked animal" (Q 3.4.95 / F 3.4.100) parallels shifts in the play between the order of nature and modes of mimesis. Lear's journey from king to beggar ultimately challenges Edmond's account of nature as *physis*, since custom and politics are responsible in part for his demise.

Languages of Nature

Edmond's manipulation of others in the play involves deceitful and theatrical representations that differ markedly from Lear's Parmenidean model of unity between language and nature. Shakespeare signals Gloucester's susceptibility to such manipulation in the latter's naïve faith in the "wisdom of Nature" as measured in eclipses and other forms of divination (Q 1.2.96 / F 1.2.95) that Gloucester takes to be indications of a misalignment in "bias of nature" leading to "ruinous disorders" and "treachery" (F 1.2.102–3 / "treason" in Q 1.2.100).[31] As we saw in the last chapter, Gloucester's embrace of astrology should not be regarded as a purely anachronistic concern that early modern audiences would have dismissed out of hand. Indeed, as Phebe Jensen notes, astrological prognostications reached a height of popularity in England in the annual almanacs published in the late sixteenth and early seventeenth centuries. Jensen links astrology in *King Lear* to important debates of early modern natural philosophy and to "deep unanswerable questions about providence, the origins of evil, the nature of the human heart, and the connections among human, nature, and divine."[32] I agree with Jensen while expanding and modifying her claims by drawing attention to the way that such questions are embedded in both theological and mimetic concerns in the play.

The widening gap between Lear's assumptions about nature and the order of nature he encounters materially generates new experiments with language, beginning with the Fool, whose questions, riddles, and songs conjure new semiotic systems and modes of theatrical representation. Unlike Cordelia and Lear, the Fool speaks from the heart, but he does not face punishment for doing so, even though he jokes to Lear that he is whipped by him for lying and by Lear's daughters for "speaking true" (Q1.4.168–9 / F 1.4.1.52–3). The Fool urges Lear to "speak less than thou knowest" (Q/F 1.4.109) and calls him an "O without a figure" (Q 1.4.177 / F 1.4.160), highlighting the growing impotence of his language and the nothingness that has begun to overtake him. Here and in other allusions to nothinginess, the Fool pushes the doctrine of *ex nihilo nihil fit* in new directions as he exposes both Lear's faults in his treatment of Cordelia and his naïve vision of language.

Like the Fool, Kent in his expressive catalogue of insults toward the Steward introduces new genres of speech in the play that challenge the systems of semiotic unity that define both Lear and Gloucester as characters. In addition to mobilizing anti-court satire in railing against the

Steward (in the Quarto text) as a "glass-gazing, superfinical rogue" (Q 2.2.16), Kent introduces modes of theatrical representation in accusing him of taking "Vanity the puppet's part" against Lear (Q / F 2.2.32). Significantly, he incorporates a key analogy to nature in his railing:

KENT	Nature disclaims in thee. A tailor made thee!
CORNWALL	Thou art a strange fellow. A tailor make a man?
KENT	A tailor, sir. A stonecutter or a painter could not have made him so ill, though they had been but two years o'th'trade. (Q/F 2.2.49–54)

Kent's verbal aggression differs in substance but not in style from the Fool. Where the Fool and Lear support the convergence of nature and artificial convention assumed by Lear, Kent emphasizes their separation, making explicit the divide between Lear and the daughters whose homes he shuttles between. As the storm scene approaches, Lear remains committed to his vision of metaphysical plenitude even as such plenitude wanes. When Goneril asks him why he needs even five followers when staying with her, given that she has many servants of her own to attend to him, Lear explodes:

> Oh, reason not the need! Our basest beggars
> Are in the poorest thing superfluous.
> Allow not nature more than nature needs,
> Man's life is cheap as beast's. Thou art a lady:
> If only to go warm were gorgeous,
> Why, nature needs not what thou gorgeous wear'st,
> Which scarcely keeps thee warm. (Q 2.2.433–9 / F 4.4.445–51)

Lear argues that clothing is "natural" because, unlike animals, humans require artificial covering or else suffer pain and even death in the face of the elements. As Laurie Shannon notes, the condition of nakedness that Lear here describes and later experiences draws on religious and theological forms of early modern natural history that articulate both "negative human exceptionalism" of animals and forms of animal "sovereignty" that reveal human dependency on animals for clothing.[33] Nature in this passage thus adopts the terms of artifice that Kent has introduced. But, as Lear descends the Great Chain of Being, such "nature" begins to lose its explicit and metaphysical theological justification. When he utters his curse against

Goneril and Regan as "unnatural hags" (Q 2.2.447 / F 2.2.459), for example, Lear does not invoke the gods as he had in earlier scenes. In the place of divine retribution, Lear promises the "terrors of the earth" (Q ln. 451 / F ln. 463) without recognizing that he himself will experience those terrors shortly. Such terrors are not only rudely material, but also expressions of central problems of classical, medieval, and early modern natural philosophy.

Lear's initial companions in the storm, Kent and the Fool, are linked both physically and symbolically with the fallen monarch they attend. Kent is a humbled figure who embraces manual labor and sacrifices his body for his former master. Like Cordelia, Kent and the Fool also both speak honestly even as they impersonate others and transform language. Initially, Lear's unease in the storm is psychological rather than physical, focused on his feelings and his inner spiritual life.[34] He begins to acknowledge his own age and frailty, urging "Nature" to "Blow winds and crack your cheeks" (Q/F 3.2.1). This direct address to nature is more descriptive than performative, as is his subsequent claim that "Nature" has cracked its "molds" (Q/F 3.2.8), indicating that he acknowledges his own distance from the divinity of nature he previously thought to control. In releasing his hold on such nature, he also invites it to become an object of inquiry rather than a source of monarchical power. Lear ascribes divine power to the storm when he defines it as justice from the gods who will find out "undivulgèd crimes" against those who merely simulate "virtue" (Q 3.2.53, 55 / F 3.2.52, 54). But such nature remains immanent rather than transcendent, and it is significant that Lear does not call on the gods again in the play. As Kent suggests, the storm tests Lear's own "nature" as a mortal man who can suffer "affliction" (Q/F 3.2.48–9) rather than the nature of those who have moved against him. The thunderous tempest even affects his body directly by colonizing his "mind" and invading his "skin" (Q 3.4.12, 7 / F 3.4.13, 8).

Questioning Nature

Lear, Kent, and Edgar cast off older forms of knowledge and identity in the wake of storm.[35] Early on, Lear focuses mainly on "filial ingratitude" from his daughters, assuming for instance that "nothing could have subdued" the "nature" of Poor Tom except for "unkind daughters" (Q 3.4.14, 62–3 / F 3.4.15, 67–8). Yet, inspired by the non-normative language of Edgar as Poor Tom, who casts himself as a former "servingman" who used to join in the forms of flattery and excessive dress associated with the court (Q lns. 74–7 / F lns. 79–83) and who now says he "eats the swimming frog" and "cow dung for salads" (Q lns. 114, 116 / F lns. 118, 120), Lear changes his tune and addresses Poor Tom as a "philosopher" with whom he can converse (Q ln. 138 / F ln. 141). For the first time in the play, Lear engages in an open-ended conversation with another person. Impersonating a natural philosopher, he asks, "What is the cause of thunder?" (Q ln. 139 / F ln. 142). This simple question does not compel language from others, but, rather, invites philosophical conversation about the *ordo creationis.* We are not privy to Edgar's answer to Lear's meteorological inquiry, and we should not assume that he provides one. But Lear's interrogative mood signals a shift in his attitude toward nature and language in the context of a classic topic of natural philosophy.[36]

In his treatise on *Meteorology*, for example, Aristotle defines thunder as a "dry exhalation" that causes earthquakes when it flows in the earth and either condenses into water in the form of rain or causes thunder as it is enveloped in a cloud (370a). But later Christian theologians viewed thunder as a manifestation of God's power that could not be understood fully using scientific methods of inquiry. For Calvin, for example, thunder caused observers of the Book of Nature to be "astounded and overwhelmed in ecstasy."[37] For La Primaudaye, moreover, thunder represents God's majesty—expressing "how terrible and fearful He showeth himself."[38] Thunder therefore exposes fault lines between early modern theology and natural philosophy, even though many accounts acknowledged both material and divine influences.[39]

Lear's philosophical question also marks a significant shift from his earlier relationship with the order of nature. In act 2, for example, Lear curses Goneril as a "carbuncle" in his "corrupted blood" and brags that he has the power to bid the "thunder-bearer shoot" her for her disobedience (Q 2.2.391–2, 394 / F 2.2.404–5, 407). Mocking this belief in the divine providence of weather, however, Edmond invokes the "thunder" of "revenging gods" against parricides in creating his false narrative to deceive Gloucester into thinking that Edgar intends to kill him ("revengive" in Q 2.1.45–6 / F 47–8). Lear's open-ended query about the cause of thunder during the storm scene thus signals more direct interest in nature as a material reality rather than as part of a theocratic and transcendental metaphysics that undergirds his power. As in previous invocations of natural philosophy, new modes of representation accompany this transition once again, beginning with Lear's mock trial of his daughters (developed most fully in the Quarto version). In this extended scene, Lear names his desire to "anatomize" Regan in order to see if there is "any cause in nature that makes these hard hearts" (Q 3.6.67–8 / F 3.6.35–7). This allusion to the nascent science of anatomy based on direct observation of the body rather than on textual authorities is unique in Shakespeare's works. As we saw in Chap. 6, the practice of anatomy was revived and transformed by Vesalius in the sixteenth century as part of an effort to test Galenic descriptions of the body examining human cadavers directly. Vesalius insists that his printed anatomy follows nature (*secundum naturam*) in the manner of Aristotle.[40] Yet, as Paula Findlen notes, the *Fabrica* of Vesalius forged a new relationship between nature and representation: "Nature had to be reconstructed within a microcosm," she notes, "creating an artificial world of knowledge in which scholars prodded, dissected, and experimented with nature in order to know it better."[41] John Banister's 1578 anatomy adapting Vesalius for English readers thus praises the anatomist for discovering "new shapes in nature" that were not present in Galen's accounts of the body, noting that direct observation of bodies in dissection provides a corrective "glass" or mirror against previous ignorance.[42]

Although some scholars continue to view the anatomy of Vesalius as a secular form of knowledge production, Vesalius clearly understood his work in religious terms and regarded the *ordo creationis* as an extension of divine power. For example, the *Fabrica* portrays the body as a wonder of divine creation in both the design of bones to create a foundational structure for the body and in particular organs. Banister echoes Vesalius in noting that it was the "mighty pleasure of God (whom we call Nature)" to create the bones without nerves because of their privileged role in supporting the body.[43] The anatomist for both Vesalius and Banister has a special obligation to uncover the structure and meaning of the human body as a product of God, working backwards in relation to His process of creation as they uncovered divine designs in bones, organs, and ligaments by carefully disassembling cadavers layer by layer.

But there was an important tension, as Sarah Keleher notes, between the divine work of the anatomist and the fallen bodies of the cadavers themselves, signified in the text of Vesalius multiple references to and images of Adam and Eve. Keleher identifies a central soteriological narrative in Vesalius that surrounds both the theological problem that the soul has fled from the deceased cadavers (only to be reunited at Judgement Day) and the medical problem that the soul in many philosophical accounts was thought to have a material presence in the body. Even though anatomists could only detect traces of the soul, they understood that it was part of an invisible body that could not be accessed through dissection. For Vesalius in particular, only the fully resurrected body could offer the full and transparent knowledge that the fallen body lacks. In pursuing a complete account of the skeleton and other bodily structures, as Keleher argues, Vesalius views his own study and printed text of the *Fabrica* as part of the resurrection narrative of the human body, disassembling cadavers in anatomy theaters in order to reassemble them in both their fallen and saved versions on the printed page.[44] This printed text functions, once again, as an integral tool of scientific inquiry that participates in modes of scientific discovery.

The imaginary trial of Goneril and Regan employs theatrical genres in ways that overlap with the work of anatomy in early modern anatomy theaters. Both function as spaces of embodied performance.[45] Like the public theater of Shakespeare and his contemporaries, the anatomy theater also invited direct observation as a mode of knowledge, even though the experience of watching a play obviously differed in important ways from that of witnessing an anatomy.[46] Where the anatomy theater focused the gaze of spectators on a single corpse, for example, playhouses such as the Rose and the Globe produced living, speaking bodies in motion on the stage. At the same time, as I have been arguing, Shakespeare's theater developed new structures of representation based on earlier traditions of drama that themselves drew their mimetic capacity, in part, from modes of natural philosophy and theology as they reframed nature in early modern England.

Lear's mock trial of his daughters occurs almost simultaneously in the play with the torture of Gloucester by Cornwall and Regan, which Cornwall authorizes on the grounds of his own "power" rather than a formal trial of "justice" (Q/F 3.7.25). Where that (mis)trial leads to the real blinding of Gloucester, Lear's trial involves only imaginary figures, including figures that Lear, Edgar, and Kent play themselves and the "joint-stool" that the Fool mistakes for Goneril (Q 3.6.44).[47] Such metatheatricality comes most fully into view in the climactic suicide attempt by Gloucester, managed by his disguised son Edgar. Gloucester's torture by Cornwall leads not just to physical blindness but also spiritual despair, while Edgar's impersonation as Poor Tom creates dramatic irony in the encounter between father and son. After leading Gloucester to the supposed cliffs of Dover, Edgar offers a detailed description of the scene for his blind father:

> Here's the place. Stand still. How fearful
> And dizzy 'tis to cast one's eyes so low!
> The crows and choughs that wing the midway air
> Show scarce so gross as beetles. Halfway down
> Hangs one that gathers samphire: dreadful trade!
> Methinks he seems no bigger than his head.
> The fishermen that walked upon the beach
> Appear like mice, and yond tall anchoring bark
> Diminished to her cock; her cock, a buoy
> Almost too small for sight. The murmuring surge,
> That on th'unnumbered idle pebble chafes,
> Cannot be heard. (Q 4.6.13–24 / F 4.6.12–23)

As several scholars have noted, Edgar's language draws on the arts of perspective to create visually spectacular language that goes beyond what Shakespeare's stage can show.[48] His words invoke the science of optics that was the subject of Chap. 6, and that likewise both draws on and challenges the arts of perspective. Henry S. Turner argues, for example, that the passage introduces a vertical horizon that gestures toward an unrepresentable kind of mathematical infinity, one that complements the form of nothingness in the "void" of the storm. Both forms of unrepresentability contribute to what Turner describes as the "dizzying mimetic complexity" of the scene and the play more broadly, a form of mimesis that is constantly engaged in negotiating the terms of its own theatrical capacity to represent space and place.[49] But Edgar's descriptive language also challenges the logic of linear perspective. The vertical plane of the cliffs has no vanishing point, for example, and most of Edgar's description focuses on the limits rather than the powers of human sight. Crows appear so small as to resemble beetles and the "cock" or cockboat towed behind a larger vessel is "almost too small for sight." Edgar's verbal description of vertical space draws attention to the sightlessness of his father while acknowledging the limits of stage representation. Even aural information is difficult to receive. In a clever ploy to convince Gloucester that he is very high above the sea that also works to convince playgoers of the reality of the scene, Edgar notes that the waves on the beach are too far away to hear distinctly.

Shakespeare's staging of the arts of perspective also serves an explicitly spiritual narrative of redemption. Edgar's careful manipulation of his father's despair aims to "cure" him of it (Q/F 4.6.36). Before he falls, Gloucester says a prayer to the "mighty gods" in which he renounces the "world" and acknowledges that it is his fate to die rather than to let his "snuff and loathèd part of nature [...] burn itself out" (Q/F 4.6.36–42). Gloucester describes himself in these lines in opposition to the gods but as part of the order of nature. For his part, instead of dissuading his father from committing suicide by invoking Christian decrees against it or using alternate arguments, Edgar utilizes mimetic modes of descriptive language and impersonation to construct an image of virtual reality. In the moments after Gloucester thinks that he has fallen from a very high cliff, Edgar impersonates a new character who supposedly discovers Gloucester and shows his amazement, as a body of "heavy substance," he could have fallen so far and yet "bleed'st not" (Q/F 4.5.51–4). Unable to account for Gloucester's fall based on his knowledge of nature, Edgar declares, "Thy life's a miracle" (Q 4.6.57). This is one of only two invocations of a miracle in the tragedy, the earlier coming when the King of France in act 1 refers to Lear's rejection of Cordelia as a "miracle" because he cannot understand it using "reason" (Q 1.1.201–10 / F. 1.1.212–21). But Edgar frames Gloucester's attempted suicide and resurrection in terms of a Christian miracle as part of the broader shift in the play from pagan to Christian metaphysics.

As Lear descends into madness, he takes on the verbal tics of the forgotten (and, as we later learn, deceased) Fool by speaking metaphorically, jumping from one topic to the next and moving metonymically across imaginative associations. The fallen king appropriately describes himself as the "natural fool of fortune" (Q 4.6.178 / F 4.5.185) and immerses himself physically and spiritually in the orders of nature. For both Lear and Gloucester, the space of nature outside the court simultaneously registers a fall into nature and acknowledges such nature as a source of self-knowledge and spiritual regeneration.[50] The barren landscape of the second half of the play brims with life and alters earlier assumptions about the nature of land. In the opening division of the kingdom, Lear conflates the landscape of the kingdom with the map on which he has divided territory, abstracting the complex material reality of the natural world into a model that can be manipulated to serve his own political interests and to force his daughters and others to obey his power.[51] Yet the void of nature contains vital powers, including the "many simples operative" that offer healing balms to Lear and that Cordelia refers to as the "unpublish'd virtues of the earth" (Q 4.4.13, 16 / F 4.3.13, 15).[52] Such "virtues" mixed with sleep (the "foster nurse of nature," as the Doctor notes) recall and draw on the language of early modern herbals such as those we examined in Chap. 2 (Q 4.4.12 / F 4.4.11). As Susan C. Staub observes, for example, the weeds gathered as a natural crown on Lear's head are "teeming with plant life," while Arthur Kinney notes that his "natural crown" is simultaneously poisonous and therapeutic.[53] The plants mentioned in the play were used, for example, to cure diseases of the brain such as epilepsy in early modern England. Unlike Friar Laurence in *Romeo and Juliet*, Lear does not catalogue the virtues of the *ordo creationis* he encounters. Yet Lear's physical and spiritual regeneration nevertheless depends on such virtues. When Cordelia recalls how her father was met by her servants, she describes him as "Crown'd with rank fumitory and furrow weeds, / With hardocks, hemlock, nettles, cuckoo flowers, / Darnel, and all the idle weeds the grow / In our sustaining corn" (Q / F 4.3.3–6). The crown of monarchy that Lear attempted to divide in the first scene of the play here returns as a natural crown of weeds that also symbolizes, at least in part, the crown of thorns of Jesus.

Like Gloucester's blindness and Poor Tom's physical transformation, Lear's spiritual *kenosis* includes the stripping away of custom and law. In such reduced circumstances, Lear attempts to construct a utopian space with Cordelia based on natural rather than supernatural foundations. Seized by enemy forces, Lear seeks to reassure her by outlining a vision of their future together:

> Come, let's away to prison.
> We two alone will sing like birds i'th' cage.
> When thou dost ask me blessing, I'll kneel down
> And ask of thee forgiveness. So we'll live,
> And pray, and sing, and tell old tales, and laugh
> At gilded butterflies, and hear poor rogues
> Talk of court news, and we'll talk with them too—
> Who loses, and who wins; who's in, who's out—
> And take upon 's the mystery of things,
> As if we were God's spies. And we'll wear out,
> In a walled prison, packs and sects of great ones,
> That ebb and flow by th' moon. (Q/F 5.3.8–19)

Lear registers his fallen condition in and through the *ordo creationis*, picturing himself and Cordelia first as caged birds and then as passive bodies in the "ebb and flow" of the tide. Lear's idealized vision combines natural and artificial features, captured most fully in the central and dissonant image of a gilded butterfly. This hybrid symbol refers on the surface to ostentatious courtly fops like Osric in *Hamlet,* but it also symbolizes the artificial mimesis at the heart of the order of nature in the play. The confined space of the prison cell that Lear idealizes is simple, resembling a monastic cell for prayer, blessing, and forgiveness. It is also a space of natural conversation, song, laughter, and "tales" that represent once more new modes of language in the play. Lear no longer pretends to command the gods or a divine nature. Rather, he is content to passively take in the world not from the omnipotent position of God but rather as one of "God's spies."

Lear's utopian vision is never realized in the play, of course, but it follows the larger pattern of shifting relations between religion, science, and mimesis that we have been tracing. Such relations are also at stake in the chivalric duel between Edmond and Edgar, as Edgar triumphs over his deceitful stepbrother by demonstrating natural skill and talent as he brings to light the deceit of Edmond. Edgar refuses to announce his name before killing Edmond, violating the norms of chivalric combat and resisting the conventional order of language in favor of his silent and embodied presence. The repeated sound of the trumpet, most likely produced by the same instrument that announced plays at the Globe and other playhouses in early modern London, creates a musical version of such elemental truthfulness that affects the senses directly, without the intercession of the gods. The nonlinguistic sonority of the trumpet rises above the semantic confusion of spoken and written speech that has been the source of both unstable metaphysics and Machiavellian manipulation in earlier scenes.[54] Like the silent body of Edgar in combat, the trumpet is an immanent and natural rather than transcendent and supernatural sign. Both, however, stand metonymically for larger systems of divine signification: the trumpet in its connection with the angels of God and Edgar as a warrior for Christ. Edmond himself acknowledges the special physical presence of Edgar when he says that his "outside looks so fair and warlike" (Q 5.3.138 / F 5.3.133), while Albany notes, following Edgar's victory, "Methought thy very gait did prophesy / A royal nobleness" (Q lns. 171–2 / F lns. 167–8). The Christian overtones of this presence emerge in Edmond's response to his brother, since the villain is now "moved" by the "speech" of the gentleman who recounts how he "revealed himself" to Lear and asked for "blessing" in relating his "pilgrimage" to save him (Q lns. 188–95 / F lns. 181–92). As these words indicate, Edmond participates in the spiritual regeneration of Lear and Gloucester.

The language of Christian blessing, revelation, and pilgrimage only makes sense in Edmond's mouth after Edmond himself has renounced his own evil and given over his view of nature as *nomos*. In keeping with this altered tone, Albany identifies "the judgement of the heavens" in the deadly display of bodies and bemoans their inability to bury the corpses religiously as "very manners urges" (Q 5.3.226–29 / F 5.3.206–9).[55]

Edgar's defeat of Edmond goes some way toward repairing the gulf between signifier and signified that accompanies Lear's fall, drawing on theatrical conventions of language and embodied action to do so. Like the physical presence of Sylvia in *Two Gentlemen of Verona* analyzed in Chap. 4, this natural presence produces a new form of divine visibility in the play in which Edgar's hand-to-hand combat enacts divine justice and produces new sources of faith that are even more trustworthy than the false miracles of divine monarchy and the superstitious embrace of astrology from earlier scenes. Edgar's embodied presence emerges from the same order of nature that shapes the divine reformation of Lear and Gloucester.

Ecomimesis

Cordelia's death likely shocked early modern viewers who expected the happy ending of traditional versions of the Lear story. Samuel Johnson famously confessed that he could not tolerate reading Shakespeare's tragedy without keeping Cordelia alive, while the Irish poet Nahum Tate rewrote the play in 1681 to allow Lear to retire happily and to give Cordelia both the crown and Edgar as a husband.[56] In Shakespeare's treatment of the story, however, the plainly dressed corpse of Cordelia returns the play to questions of representing the order of nature that we have been tracing. The death of Cordelia may be Christ-like, and Lear's act of holding her may produce an inverted pietà on the stage, as Katharine Goodman has noted, but the simple presence of bodies and basic authority of utterance also deconstructs Lear's earlier faith in a transcendental order of pagan theocracy as part of the play's shift toward a Christian order of nature.[57] Despite his own fear that she is "dead as earth," Lear seeks direct and empirical knowledge of his daughter's condition by asking for a mirror to see if her "breath will mist or stain the stone" (Q 5.3.257–8 / F 5.3.235–6).[58] Wanting to believe beyond the evidence that she is still alive ("This feather stirs" [Q ln. 261 / F ln. 239]), Lear dies urging others to observe his daughter's lips in search of breath, noting that he might have saved her if he were younger. Lear claims that his younger self would have used his sword to make her killer "skip" away, but in his old age "these same crosses spoil me" (Q lns. 273–5 / F lns. 251–2). The literal referent for such "crosses" is the sword he imagines using, but symbolically the cross is, of course, a central emblem of Christian theology. Like Jesus, for example, Lear's life has been "spoiled" by the cross, but also like Christ his death leads to forms of spiritual renewal that connect him with other characters and with those who witness his story.

In offering a restorative model of justice at the conclusion of the play, Albany draws out these spiritual implications in ways that reproduce the larger shift in the tragedy from transcendental pagan *logos* to a more immanent and material order of Christian nature:

> You lords and noble friends, know our intent:
> What comfort to this great decay may come
> Shall be applied. For us, we will resign
> During the life of this old majesty
> To him our absolute power; [*to* EDGAR] you to your rights,
> With boot and such addition as your honors
> Have more than merited. All friends shall
> Taste the wages of their virtue and all foes
> The cup of their deservings. (Q 5.3.292–300 / F 5.3.272–80)

Albany's vision of the future recalls chapter 2 of Paul's Epistle to the Romans, where Paul notes that God will judge "every man according to his works" (2:6), distributing glory and eternal life to those who "do good" (2:10) and wrath to those who "are contentious and disobey the truth" (2:8). Albany's reference to "wages of virtue" also invokes and inverts the biblical allusion in Romans 6:23 ("For the wages of sin is death, but the gift of life is eternal life in Christ") as he seeks to transform the death and corruption he witnesses by promoting Edgar as the new ruler. Yet Lear's final question of why "a dog, a horse, [or] a rat" should "have life" in the wake of his own daughter's death undercuts Albany's Christian vision of reparation (Q 5.3.302 / F 5.3.282). The play remains uncertain about which vision triumphs and whether any vision can compensate for the vast death and destruction it portrays. The promise of redemption remains more implicit than explicit in the tragedy, produced, as we have seen, through shifting modes of theology, natural philosophy, and theatrical mimesis.

The end of *King Lear* also revives questions about the relation between verbal representation and ethics. Edgar's final moralizing line, "Speak what we feel, not what we ought to say" (Q 5.3.320 / F 5.3.300), sums up the play in a way that seems woefully inadequate to the human suffering it has displayed. But his words reflect the movement we have been tracing in the play from metaphysical unity of language and the order of nature to modes of Christian materialism. The line also shores up Edgar's identity, grounded in as it is in an immanent metaphysics of language and identity on which he hopes to ground the future body politic. Edgar's rejection of *nomos* and convention in favor of a vision of language as the authentic expression of interiority does not necessarily embrace the *physis* of Edmond and others, since it seeks to locate a form of authentic and basic truth in communication that leaves no room for deceit. Rather, Shakespeare situates embodied performance as a measure of human "presence" in a way that illustrates the larger structure at work in the play, one that moves away from coerced speech in the service of transcendent metaphysics to one that grounds language and representation on natural human behavior.

Shakespeare's version of the Lear story thus complicates the explicit Christian allegory of the earlier *Leir* play. Rejecting any miraculous and happy ending, Shakespeare nevertheless weaves Christian theology indirectly into the fabric of the tragedy by way of questions about the order of nature and its capacity to signify. In the process, religion is neither evacuated, as Halpern claims, nor "emptied out" through appropriation, as Stephen Greenblatt argues in a famous new historicist reading of the play.[59] Rather, as we have seen, the material world to which numerous critics have drawn attention is thoroughly imbued with divine potential. *King Lear* is thus environmentally conscious, in a sense, but it also invokes only to reject what Timothy Morton calls "ecomimesis" or "illusion of false immediacy" that would bridge the gap between subject and object and between humans and nature.[60] In a formal level, perhaps, the play does foreground ecomimesis by approaching nature in the manner of poets who seek to understand it as a form of representation.[61] To make such a claim, however, is not necessarily to endorse easy moral conclusions. Shakespeare's great tragedy ultimately favors the interrogative rather than declarative mood. Still, as this chapter has demonstrated, the proliferation of forms of nature in the tragedy breaks down conventional oppositions between material and spiritual readings. The divine order of nature is a central source of the play's materialism, which evacuates neither religion nor spirituality, joining other works by Shakespeare we have considered above that negotiate relations between early modern science and religion in and through shifting modes of theatrical mimesis.

Notes

1. On these multiple modes of nature in *Lear*, see John F. Danby, *Shakespeare's Doctrine of Nature* and Graham Martin and Stephen Regan's chapter on the play in *Shakespeare: Texts and Contexts*, ed. Ryan, 241–76, esp. 262–5. Quotations from the play are taken from the Folio text, but I include Quarto act, scene, and line numbers for reference and designate passages that only appear in one or the other text accordingly. I quote mainly from the Folio text, adding brackets to indicate significant variations in the Quarto where necessary.
2. A.C. Bradley, *Shakespearean Tragedy*, 248.
3. Ibid., 247, n. 29.
4. G.K. Hunter, *Dramatic Identities and Cultural Tradition*, 252.
5. G. Wilson Knight, *Shakespeare and Religion*, 144 et. passim. See also Robert Speaight, *Nature in Shakespearean Tragedy*, 129.
6. A.D. Nuttall, *Shakespeare the Thinker*, 312.
7. Jan Kott, *Shakespeare our Contemporary*, trans. Taborski, 116.
8. My reference to "vibrant materialism" draws on Jane Bennett's *Vibrant Matter: A Political Ecology of Things.*
9. Jonathan Dollimore, *Radical Tragedy: Religion, Ideology, and Power in the Drama of Shakespeare and His Contemporaries*, 191.
10. Richard Halpern, *The Poetics of Primitive Accumulation*, 222. Harold Bloom also locates strands of nihilism in the play in *Shakespeare: The Invention of the Human*, 292–3.
11. Stanley Cavell describes the setting of *Lear* as a "dissociated world" in an essay on *Othello*. See "Epistemology and Tragedy: A Reading of *Othello*," 31. I also disagree with William H. Steffen in his recent claim that the play favors a "secular model of redemption" that elevates the "self-correcting imbalances of nature" as a solution to "economic imbalances between kingly pomp and houseless poverty." See Steffen, *Anthropocene Theater and the Shakespearean Stage*, 187.
12. Mary Thomas Crane, *Losing Touch with Nature,* 137.
13. Michael Witmore, *Shakespearean Metaphysics*, 1.
14. William R. Elton, "*King Lear*" *and the Gods*, 9. Hereafter cited parenthetically in the text by page number.
15. Giorgio Agamben, *Homo Sacer: Sovereign Power and Bare Life*, trans. Heller-Roazen, 183.
16. The Quarto text of the line does not contain the word "nature." It reads simply "where merit doth most challenge it."
17. The Quarto text is "Nothing can come of nothing," while the Folio text has "Nothing will come of nothing."

18. Drawing on Dent, most editors regard the line as "proverbial," a phrase Stanley Wells uses in his Oxford edition of the play (104) and that R.A. Foakes repeats in his Arden 3 edition (164, n. 90). The Norton edition edited by Stephen Greenblatt includes a longer gloss: "*Ex nihilo nihil fit*" is a "maxim derived from Aristotle [...] accepted by the Christian Middle Ages with the single exception of God having created the world out of nothing" (2334, n. 1). T.W. Baldwin locates a similar phrase in a satire by the classical Roman poet Aulus Persius Flaccus ("Persius"), whose works apparently were often paired with Juvenal in early modern curricula in *William Shakespeare's Small Latin and Less Greek,* 2:542–3. Peter Milward links Cordelia's "nothing" with the nothingness from which God created the universe, citing Job 26:7 and claims that Lear's response is "merely Aristotelian," overlooking the more complex implications of the phrase I am tracing. See *Biblical Influences in Shakespeare's Great Tragedies*, 160.
19. Shakespeare may have been familiar with the proverb and with Parmenides from North's translation of Plutarch that he used in writing several plays, including *Julius Caesar* and *Antony and Cleopatra.* He may also have known a later version of the concept as incorporated into Aristotle's *Physics.* On the enduring influence of Parmenides, including his impact on Hegel and later philosophy, see Bertrand Russell, "The Problem of Infinity Considered Historically," reprinted in Russell, *Our Knowledge of the External World as a Field in Modern Philosophy*125–47, esp. 134–5; and Raymond Tallis, *The Enduring Significance of Parmenides: Unthinkable Thought.*
20. J.E. Boodin paraphrases *ex nihilo nihil fit* thus: "In order for knowledge to be possible, there must be constancy. If there is no constancy there can be no reference, no discourse." See "The Vision of Parmenides," 580.
21. See Parmenides, *The Testimonia,* part 2 (Doctrine D). The Loeb translation reads "This is the same: to think and the thought that 'is'" (ταὐτὸν δ' ἐστὶ νοεῖν τε καὶ οὕνεκεν ἔστι νόημα). In *Early Greek Philosophy*, ed. and trans. Laks and Most, 46–7.
22. See, e.g., William Franke's brief account of the significance of Parmenides in *On the Universality of What is Not*, 360.
23. Aristotle's ultimate cosmology remains somewhat unclear, and there are differences in its elaboration in books 2 and 8 of the *Physics.* For an overview of the problems, together with a summary of Aristotle's relation to Parmenides and other earlier Greek natural philosophers, see Daniel W. Graham's "Introduction" to *Aristotle's Physics*, esp. x–xvii.
24. Halpern draws on economic ideas to describe a "divorce between the signs and the material realities of royal power" in the play, which he argues

undermines the divine monarchy of King James I. My reading suggests, by contrast, that the play is more invested in theological and scientific questions than economic ones. See *The Poetics of Primitive Accumulation*, 220, 234.

25. Bradley, *Shakespearean Tragedy*, 261. See also Turner, "*King Lear* Without: The Heath," 161.
26. On this point, see Paul Cantor, "Nature and Convention in *King Lear*," 213–34.
27. On the relation between nature and custom in the play, see Mark Allen McDonald, *Shakespeare's "King Lear" with "The Tempest,"* 15–19.
28. Leonard Wright, *A Display of Duty* (1589), 1. Arthur F. Kinney alludes to this text in his discussion of the play in "Lear," 685–6.
29. The quarto text reads "dis-utered torment," which the Norton editors gloss as "perverse and unnatural." See *King Lear* (Q 1.4.266) in *The Norton Shakespeare,* ed. Greenblatt et al., 2370.
30. For a reading of these lines in relation to larger questions about the status of "natural law" in the play, see R.V. Young, "Hope and Despair in *King Lear*," in *New Critical Essays*, ed. Kahan, 264.
31. For a different account of Edmund's relationship to early modern debates about nature as law (*nomos*) versus nature as a physical and material life force (*physis*), see Robert J. Bauer, "Despite of Mine Own Nature: Edmund and the Orders, Cosmic and Moral."
32. Jensen, "Causes in Nature," 210.
33. See Shannon, "Poor, Bare, Forked," esp. 175–6.
34. As Rolf Soellner notes, "Lear before the storm stresses the physical above the psychic and spiritual elements. It is part of Lear's education through suffering that he comes to feel on his own person how much worse psychic illness is than physical pain." See "Shakespeare, Aristotle, Plato, and the Soul," 68.
35. Cantor notes in his essay on "Nature and Justice in *King Lear*" that, "like Lear, Edgar must be thrust out of the comfort of his conventional role and into the world of nature in order for him to gain insight into himself and humanity" (245).
36. Elton reads the thunder in terms of Lear's shifting views of providence and his "attitude to mankind" in *King Lear and the Gods*, 198. Danby reads the thunder in *King Lear* as a supernatural force that signifies Lear's "rehabilitation" in *Shakespeare's Doctrine of Nature*, 185. As Cantor notes, Lear's question about the source of thunder also echoes the dialogue between Socrates and the rustic Strepsiades in *The Clouds* (423 BCE) by Aristophanes, where the latter rejects the physical, material account of weather provided by Socrates because it ignores the power of the divine in shaping the social order. Cantor invokes the passage as part of his understanding of the rela-

tion between nature and justice in the play in "Nature and Justice in *King Lear*," 231.

37. Calvin, from his commentary on the Psalms, trans. Bouwsma, *John Calvin*, 164.
38. La Primaudaye, trans. R. Dolman, 204.
39. Seventeenth-century natural philosophers favored material theories of thunder, including the German atomist and alchemist Daniel Sennert, who proposed that thunder was caused by a chemical reaction similar to that caused by the explosion of gunpowder. A translation of Sennert's *Thirteen Books of Natural Philosohpy* was made, in part, by Nicholas Culpepper and published in London in 1660 by Peter Cole. The discussion of thunder is on page 108 of that volume.
40. Kusukawa, *Picturing the Book of Nature*, 217–18.
41. Findlen, "Anatomy theaters, Botanical Gardens, and Natural History Collections," 272.
42. John Banister, *The History of Man, Sucked from the Sap of the Most Approved Anatomists*, sig. Aa4.
43. Ibid., sigs. C1-C1v.
44. Sarah Keleher, "Anatomy and Salvation History in Andreas Vesalius's *De humani corporis fabrica*," in *Rethinking Science and Religion in Early Modern Culture*, ed. Rust and Kitch (forthcoming).
45. On this topic, see Hillary M. Nunn, *Staging Anatomies.*
46. See Peter G.F. Eversmann, "What Did They See? Science and Religion in the Anatomical Theatres of the Sixteenth and Seventeenth Centuries," in *Quid est Secretum*, ed. Dekonicnk et al., 260–84.
47. Gary Taylor's question "Does the joint-stool represent a political victim condemned *in absentia* by a court the plaintiff himself appoints, or does it represent instead a pathetic substitute, the closest political impotence can get to punishing its oppressors," captures the ambiguous nature of the justice produced by the trial scene as well as its problematic status as a form of theatrical representation. See "Monopolies, Show Trials, Disaster, and Invasion," in *The Division of the Kingdoms*, ed. Taylor and Warren, 92.
48. See., e.g., Jonathan Goldberg, "Perspectives," and Stephen Orgel, "Shakespeare Imagines a Theater."
49. Turner, "*King Lear* Without," 170, 177.
50. I disagree with the claim by Robert Bechtold Heilman that Lear's "sinking into insanity" offers a "means of relief from the intolerable burden of an enigmatic world." See *This Great Stage*, 192. On the misleading convention of referring to the landscape in the play as a "heath," see ibid., esp. 163–4.
51. As Dan Brayton observes, "the baldness of Lear's reduction of his daughters to tokens of exchange is matched by the evacuation of the social from

Lear's map-reading, the erasure of any indication of human activity from Lear's map of stage." See "Angling in the Lake of Darkness," 406.

52. Rebecca Bushnell includes this example in her survey of "Vegetable Virtues" in *Shakespeare's Botanical Imagination*, ed. Staub, 52–3.
53. Susan C. Staub, "The 'Idle Weeds that Grow in Sustaining Corn,'" in *Shakespeare's Botanical Imagination*, ed. Staub, 68. See also Kinney, 702.
54. Stanley Wells makes a similar point in *Shakespeare and Co.*, 7.
55. The Quarto reads "justice" rather than "judgement" of the heavens.
56. Nahum Tate, *The History of King Lear*. For a classic account of this afterlife of the play, see Maynard Mack, *Shakespeare in Our Time*, 1–45. For more recent accounts, see C.B. Hardman, "Our Drooping Country Now Erects Her Head," and Sonia Massai, "Nahum Tate's Revision of Shakespeare's *King Lears*."
57. See Katharine Goodman, "Inverting the Pietà in *King Lear*." As R.V. Young notes, however, "Cordelia is not Christ. She is a pagan woman who leads a French army into England, is defeated and captured, and brutally put to death by her captor. She will not rise from her grave in three days." See "Hope and Despair in *King Lear*," 273.
58. In Mary Thomas Crane's trenchant reading of this moment, "The densest matter seems here to be the most substantial, although Lear dies yearning for visible proof beyond it." See *Losing Touch with Nature*, 144.
59. Greenblatt, "Shakespeare and the Exorcists," in *Shakespearean Negotiations*, esp. 119.
60. Morton, *Ecology without Nature*, 182.
61. Ibid., 77.

CHAPTER 9

Epilogue

Shakespeare responded to shifting relations between science, religion, and mimesis in early modern Europe by employing new modes of theater whose ambitious scope was captured in the name of the playhouse where many of his plays were performed: the Globe. As this book has argued, Shakespeare's relationship to "nature" was multi-faceted and went beyond pastoral nostalgia for a lost green world. Yet the attraction of a post-Romantic model of nostalgia remains strong for many readers and critics. In his conclusion to *Back to Nature,* for example, Robert N. Watson cites approvingly Ted Hughes's claim that the story of "Western man [*sic.*]" is that of "the mind exiled from Nature."[1] Watson argues that the "terrified collective mind of Western Europe" clung "tighter than ever to the outer face of the natural world, knowing that the surface is slippery and crumbling, and knowing that a fall may do irreparable damage."[2] This study challenges Watson's notion of nostalgia by analyzing a more complex and historically embedded account of nature as a dynamic and hybrid response to classical, medieval, and early modern debates in natural philosophy and theology. Such nature, as we have seen, was also increasingly dependent on and embedded in forms of mimesis. Early modern science, religion, and modes of theatrical representation worked together to uncover and expand forms of representation within the order of nature, which themselves spurred new inquiries into that order.

A. Kitch, *Shakespeare's Theater of Nature*, Palgrave Studies in Literature, Science and Medicine,
https://doi.org/10.1007/978-3-031-78082-0_9

This is not to say that early modern science and religion were reducible to one another. For example, some forms of early modern natural science rejected Christianity explicitly, while others used religion sparingly or only to justify their right to exist. As Stephen Gaukroger notes, moreover, new conflict emerged in Europe as Christianity "took over natural philosophy in the seventeenth century, setting its agenda … and in the end establishing it as something in part constructed in the image of religion."[3] This argument reflects the overall claim of this book that early modern efforts to understand nature were neither secular nor teleological. But this does not mean that they were free from ideology. As Lorraine Daston argues in her brief but spirited manifesto *Against Nature,* any effort to construe a "uniform and inviolable order" of nature is inevitably fraught, especially when it attempts to ground moral values on ideas of nature.[4] Daston thus bemoans the historical justifications of activities such as breastfeeding and slavery as natural (6). Her objections are both moral and metaphysical, as she notes that many invocations of nature fail to capture the "embarrassingly rich" varieties and modes of nature. According to Daston, before nature underwrote a "secular metaphysics" during the Enlightenment, it was far less orderly (31). Yet she also offers two counterexamples to this history from the seventeenth century: Robert Boyle's analogy of the laws of nature with the Strasbourg clock (27–29) and the rise of the curiosity cabinet or *Wunderkammer* among aristocratic collectors. However, even these examples fail for Daston to account for the early modern "polyphony of nature":

> The polyphony of nature is […] precisely the point: it is difficult—perhaps impossible—to imagine an order that is not manifestly, flamboyantly on display in nature. Nature is that delightful paradox, a disorderly *Wunderkammer* of all possible orders. The Renaissance *Wunderkammer*, ancestor of modern museums, dramatized the fecundity and plenitude of nature in its floor-to-ceiling displays, juxtaposing flies in amber and stuffed crocodiles, two-headed cats and striped tulips, magnets and petrified wood in order to overwhelm the spectator with the glorious hodge-podgery of it all. (61)

The *Wunderkammer* triumphs for Daston precisely because it reflects an order of nature that is flamboyantly "on display." Like Paula Findlen, who investigates the scientific importance of "jokes of nature" such as shells and other objects that take on human characteristics, Daston regards the order of nature as an archive of raw imaginative potential that can never be fully taxonomized.[5] Daston's description of nature as a structure of representation that changes over time aligns with my own account in several ways. Like the *Wunderkammer*, Shakespeare's nature is complex and capricious, drawing on traditions of natural philosophy and theology without being confined by them or trying to parallel them as truth claims. His plays also delight in the "hodge-podgery" of nature while revealing a now-familiar paradox concerning efforts to return to nature in the period: the more the order of nature becomes an object of direct observation, the more it turns out to be structured as a form of representation. Prominent natural philosophers such as Francis Bacon and Robert Boyle argued that nature both incorporated artificial forms of manufacture within its processes and that there were underlying similarities between the artificial and the natural as categories.[6] Over the course of the seventeenth century, as in later centuries, such artifice was increasingly understood as part of the fabric of nature itself. Against scholastic views of the impossibility of artificial life, for example, Descartes emphasized the power of simulation. His claim that humans without souls are mechanical structures empties the Book of Nature of its theological force, even though he also attempts to justify his natural philosophy in Christian terms.[7] And, as Steven Shapin and Simon Schaffer note in *Leviathan and the Air Pump*, experimental scientists of the Royal Society created diagrams with visually arresting but inaccurate illustrations that helped to convince audiences of the authenticity of the empirical method.[8]

Although Daston acknowledges that the formulation of natural laws in the seventeenth century was "as much theological as natural philosophical" (29), she mostly brackets religious questions. As this study has demonstrated, however, early modern nature was imbued with modes of divinity, including what Cordelia refers to as the "unpublished virtues" of the earth in *King Lear*. Shakespeare's dramatic art reveals that the order of nature was a fertile ground for theatrical experiments that might extend its mimetic capacity. As formal expressions of shifting relations between early modern science, religion, and mimesis, moreover, his plays anticipate larger epistemic shifts in the seventeenth century and beyond.

For Richard Rorty, these early modern shifts in understanding nature signaled very bad news for philosophy by positing a stark divide between humans and the natural world, with damaging implications for subsequent philosophers. Rorty laments the "glassy essence" of Cartesian metaphysics that, he claims, has alienated us from our own nature:

> It is glassy—mirror-like—for two reasons. First, it takes on new forms without being changed—but intellectual forms, rather than sensible ones as material mirrors do. Second, mirrors are made of a substance which is purer, finer grained, more subtle, and more delicate than most [...] something we share with the angels, even though they weep for our ignorance of its nature.[9]

According to Rorty, fields of philosophy from epistemology to hermeneutics have been contaminated by the desire for such "glassy mirrors" ever since Descartes. To counter such alienation, he advocates non-systematic philosophies "without mirrors" that he identifies in the writing of three "great, edifying, peripheral" figures in the history of philosophy: Martin Heidegger, John Dewy, and Ludwig Wittgenstein.[10] In his sweeping account of the history of philosophy, however, Rorty offers an inadequate picture of early modern intellectual life, failing for example to acknowledge how authors such as Shakespeare and other artists respond to and shape such life. Hamlet's Mirror of Nature is not Rorty's divided mirror of philosophy. Shakespeare's theater of nature does not produce a systematic world picture, nor does it reduce ways of knowing to mental states. Rather, as we have seen, his plays explore the semiotic plenitude of nature as a signifying system.

Shakespeare's foregrounding of nature as a scientific, mimetic, and religious principle also produces what Anthony Dawson calls an "indirect and diffuse" form of theology.[11] This indirect approach separated Shakespeare from many of his peers, including Christopher Marlowe, who made religious conflict central to his plays and embraced numerous heterodoxies in his own life.[12] Ben Jonson also courted punishment in his representation of polemical topics, one of which landed him in jail and earned him a brand of "T" (for "Tyborn" jail) on his thumb. John Webster and Thomas Middleton likewise cut their theatrical teeth by staging religious controversy more directly, including the most popular play of Jacobean England, Middleton's *Game of Chess* (1624), which satirized the failed marriage negotiations between Prince Charles and his intended Catholic bride, the Spanish Infanta Maria Anna. Shakespeare, by contrast, seems to have preferred religious representation grounded in philosophical debates that drew on ideas from theology and natural philosophy rather than stage such polarizing and dangerous topics overtly, even in the plays he wrote before the Blasphemy Law of 1606 outlawed the direct representation of God on stage so as to end "the abuse of the Holy Name of God in Stage-plays."[13]

Shakespeare's indirect theology can also be understood in terms of the history of secularism such as that proposed by Charles Taylor in *A Secular Age* (2007). Taylor refines the "disenchantment" theory of Max Weber by drawing attention to the ways that religious beliefs were reconfigured rather than simply rejected in the transition from a pre-modern to a modern world. He describes the capacious and complex modes of divinity in early modern and pre-modern societies, not just in public and political spheres but also in everyday beliefs and practices. For example, Taylor notes that, "in the pre-modern world, meanings are not only in minds, but can reside in things, or in various kinds of extra-human and intra-cosmic subjects."[14] He also identifies new concepts of the "autonomy of nature" beginning in the thirteenth century and stimulated by the introduction of new translations of Aristotle. For Taylor, such autonomy emerged from twin concepts of the divine—both "devotion to God as the creator of an ordered cosmos" and a new "evangelical turning to the world" that recognizes the presence of Christ "among the people."[15]

Taylor's dialectical view of Christianity resists linear narratives of secularization by recognizing the religious motivations at work in apparently secularizing forces. This book locates additional relations between early modern religion, natural philosophy, and mimesis that reinforce Taylor's claims. In weaving classical, medieval, and early modern philosophical and theological debates about the order of nature into his plays, Shakespeare participated in new and shifting relations between science and religion. His rich and varied works of drama also generated new modes of representation that were often revealed as central to the order of nature itself.

Notes

1. The quote by Hughes appears in his review of Max Nicholson's *Environmental Revolution* in an essay for *The Spectator* entitled "An Idea Whose Time Has Come," 378. Watson quotes him in *Back to Nature,* 324.
2. Watson, *Back to Nature,* 324.
3. Stephen Gaukroger, *The Emergence of a Scientific Culture*, 23.
4. Daston, *Against Nature*, 23. Hereafter cited parenthetically in the text by page number.
5. Findlen, "Jokes of Nature and Jokes of Knowledge."

6. Newman, *Promethean Ambitions*, esp. 271–89. See also Principe, *The Aspiring Adept.*
7. Descartes grounded his mechanical philosophy on his theological exploration, including his reading of Aquinas. See, e.g., Deborah J. Brown and Calvin G. Normore, *Descartes and the Ontology of Everyday Life*, 7. For a useful overview of the significance of automata to his philosophy, see 63–93 of that study.
8. Steven Shapin and Simon Schaffer, *Leviathan and the Air Pump*, esp. chapters 2 and 6.
9. Richard Rorty, *Philosophy and the Mirror of Nature*, 63.
10. Ibid., 368.
11. Anthony Dawson, "The Secular Theater," 240.
12. Even without assuming the truth of the various charges against Marlowe in the so-called "Baines libel," including accusations of sodomy and atheism, Marlowe's plays and biography constantly placed him in danger, as David Riggs outlines in *The World of Christopher Marlowe.*
13. *Statutes of the Realm*, 4:1097. For a longer view of this statute in relation to efforts to stage God, see Sarah Beckwith's chapter on "The Gift of God for the People of God," 161–90.
14. Charles Taylor, *A Secular Age*, 33.
15. Ibid., 94.

Works Cited

Primary Sources

Agrippa, Heinrich Cornelius. *Opera*. 2 vols. Lyon, 1600.

Alberti, Leon Battista. *On Painting*. Trans. Rocco Sinisgalli. Cambridge: Cambridge University Press, 2011.

Anon. *Here Beginneth the Seeing of Urines*. London, 1525.

Anon. *The History of the Damnable Life and Deserved Death of Doctor John Faustus* (1592). Ed. William Rose. London: Routledge, 1925.

Anon. *Towneley Creation and the Fall of the Angels* (c. 1400-1450). Ed. Garrett J.P. Epp. University of Rochester Library. https://d.lib.rochester.edu/teams/text/epp-the-creation.

Aquinas, Thomas. *Summa Theologica: Volume 4 (Part III, First Section)*. New York: Cosmio, 2007.

Aristotle. *Complete Works*. 2 vols. Ed. Jonathan Barnes. Princeton: Princeton University Press, 1984.

Aristotle. *Physics: Book VIII*. Trans. Daniel W. Graham. Oxford: Clarendon Press, 1999.

Augustine of Hippo. *Augustine on the Psalms*. Ed. and Trans. Scholastica Hebgin and Felicitas Corrigan. 2 vols. Westminster, MD: Newman Press, 1960–61.

Augustine of Hippo. *Augustine: The Writings Against the Manichaeans, and Against the Donatists*. In *Nicene and Post-Nicene Fathers*. Ed. Philip Schaff. 14 vols. New York: W.R. Eerdmans, 1956–2022.

Augustine of Hippo. *The Literal Meaning of Genesis*. Trans. John H. Taylor. New York: Newman Press, 1982.

A. Kitch, *Shakespeare's Theater of Nature*, Palgrave Studies in Literature, Science and Medicine,
https://doi.org/10.1007/978-3-031-78082-0

Bacon, Francis. *Instauratio Magna*. Ed. Graham Rees and Maria Wakely. Oxford: Oxford University Press, 2004.

Bacon, Francis. *Novum Organum*. Ed. Lisa Jardine and Michael Silverthorne. Cambridge: Cambridge University Press, 2000.

Bacon, Francis. *The Advancement and Proficience of Learning*. London, 1640

Bacon, Francis. *The Advancement of Learning*. Ed. Michael Kiernan. Oxford: Oxford University Press, 2000.

Bacon, Francis. *The Works of Francis Bacon*. Ed. James Spedding, Robert L. Ellis, and Douglas D. Heath. 15 vols. Cambridge: Cambridge University Press, 1857-64.

Bacon, Roger. *Roger Bacon's Philosophy of Nature: A Critical Edition, with English Translation, Introduction, and Notes, of "De multiplicatione specierum" and "De speculis comburentibus."* Ed. David C. Lindberg. Oxford: Clarendon, 1983.

Bale, John. *A New Comedy or Interlude Concerning the Three Laws of Nature*. London, 1562.

Banister, John. *The History of Man, Sucked from the Sap of the Most Approved Anatomists*. London, 1578.

Batman, Stephen. *The New Arrival of the Three Graces*. London, 1581.

Bodin, Jean. *Universae naturae theatrum*. Lyon, 1596.

Bodleian MS Ashmole 208. https://shakespearedocumented.folger.edu/resource/document/formans-account-seeing-plays-globe-macbeth-cymbeline-winters-tale.

Borlik, Todd Andrew, ed. *Literature and Nature in the English Renaissance: An Ecocritical Anthology*. Cambridge: Cambridge University Press, 2019.

Boyle, Robert. *Some Considerations Touching the Usefulness of Experimental Natural Philosophy*. Oxford, 1663.

Boyle, Robert. *Works*. Ed. Edward Bradford Davis, Michael Cyril, and William Hunter. 14 vols. London: Pickering & Chatto, 1999–2000.

Brahe, Tycho. *Opera Omnia*. Ed. John Louis Emil Dreyer. 15 vols. Copenhagen: Nielsen & Lydiche, 1913–1929

Browne, Thomas. *Pseudodoxia epidemica*. Ed. Kevin Killeen. Oxford: Oxford University Press, 2014.

Bullein, William. *Bulleins Bulwarke of Defense Against all Sickness*. London, 1576.

Bullinger, Henry. *An Exhortation to the Ministers of God's Word*. Trans. John Cox. London, 1575

Calvin, John. *The Institution of Christian Religion*. Trans. Thomas Norton. London, 1561.

Calvin, Jean. *Institutio Christianae religionis*. London, 1576.

Calvin, John. *Calvin's Commentaries*. Trans. John W. Fraser. 10 vols. Ed. David W. Torrance and Thomas F. Torrance. Grand Rapids, Mich.: William B. Eerdmans, 1960–1972.

Calvin, John. *Institutes of the Christian Religion*. Trans. Henry Beveridge. 2 vols. Grand Rapids, MI: WM. B. Eerdmans, 1979.

Camerarius, Joachim. *History of Strange Wonders*. London, 1561
Cavendish, Margaret. *The Blazing World and Other Writings*. Ed. Kate Lilley. London: Penguin, 1992.
Caxton, William. *Here Beginneth the Book Called the Mirror of the World*. London, 1481.
Chambers, E.K. *The Elizabethan Stage*. 4 vols. Oxford: Clarendon, 1923.
Chambers, E.K. *The Medieval Stage*. 2 vols. Oxford: Oxford University Press, 1903.
Cicero, Marcus Tullius. *"The Nature of the Gods" and "On Divination."* Trans. C.D. Yonge. Amherst, NY: Prometheus Books, 1997.
Cicero, Marcus Tullius. *De Divinatione*. Trans. W.A. Falconer. Cambridge, MA: Harvard University Press, 1923.
Clever, William. *The Flower of Physic*. London, 1590.
Cook, G.H., ed. *Letters to Cromwell and Others on the Suppression of the Monasteries*. London: John Baker, 1965.
Copernicus, Nicholas. *De Revolutionibus*. Ed. Herbert Maria Nobis and Bernhard Sticker. Hildesheim: Gerstenberg Berlag, 1984.
Culpepper, Nicolas. *The English Physician, or An Astrologo-physical Discourse of the Vulgar Herbs of this Nation*. London, 1652.
Cuningham, William. *The Cosmographical Glass, Containing the Pleasant Principles of Cosmography, Geography, Hydrography, or Navigation*. London, 1559.
Digges, Leonard. *A Geometrical Practice, Named Pantometria*. London, 1571.
Dubreuil, John. *Perspective Practical, or, A Plain and Easy Method of True and Lively Representing All Things to the Eye at a Distance by the Exact Rules of Art*. Trans. Robert Pricke. London, 1672.
Dümmler, Ernst, Ludwig Traube, Paul von Winterfeld, and Karl Strecker, eds. *Monumenta Germaniae Historica*. Berlin, 1884.
Elizabeth I, Queen. *The Second Tome of Homilies*. London, 1563.
Erasmus, Desiderius. *Erasmus: The Education of a Christian Prince, With the Panegyric for Archduke Philip of Austria*. Ed. Lisa Jardine. Cambridge: Cambridge University Press, 1997.
Erasmus, Desiderius. *The Complaint of Peace*. Trans. Thomas Paynell. London, 1559.
Euclid. *The Elements of Geometry*. Trans. H. Billingsley. Ed. John Dee. London, 1570.
Farmer, John S., ed. *Recently Recovered "Lost" Tudor Plays*. New York: Barnes and Noble, 1966.
Forman, Simon. *The Grounds of Longitude*. London, 1591.
Foxe, John. *Acts and Monuments*. London, 1563.
Gale, Thomas. "A Brief Declaration of the Worthy Art of Medicine." In *Certain Works of Galen*, sigs. A5v-A6. Trans. Gale. London, 1586.
Galen. *Certain works of Galen, Called "Methodus medendi," With a Brief Declaration of the Worthy Art of Medicine*. Trans. Thomas Gale. London, 1586.
Galen. *On the Usefulness of the Parts of the Body (De usu partium)*. 2 vols. Trans. Margaret Tallmadge May. Ithaca: Cornell University Press, 1968.

Galilei, Galileo. *Discourse on Bodies in Water.* Ed. Drake Stillman. Trans. Thomas Salusbury. Urbana: University of Illinois Press, 1960.

Galilei, Galileo. *Discoveries and Opinions of Galileo.* Ed. and Trans. Drake Stillman. New York: Random House, 1957.

Galilei, Galileo. *Sidereus Nuncius, Or, the Sidereal Messenger.* Ed. and Trans. Albert Van Helden. Chicago: University of Chicago Press, 1989.

Geveren, Sheltco à. *Of the End of This World and the Second Coming of Christ.* Trans. Thomas Rogers. London, 1577.

Gerard, John. *The Herbal, or General History of Plants.* London, 1597, 1633.

Gorham, George Cornelius. *Gleanings of a Few Scattered Ears During the Period of the Reformation in England and of the Times Immediately Succeeding.* London, 1857.

Gosson, Stephen. *Plays Confuted in Five Actions.* London, 1582.

Greg, W.W. *A Bibliography of the English Printed Drama to the Restoration.* 4 vols. London: Bibliographical Society, 1939–59

Gundisalvo, Domingo. *On the Division of Philosophy.* Trans. M. Clagett and Edward Grant. In *A Source Book in Medieval Science,* 59-76. Ed. Edward Grant. Cambridge, MA: Harvard University Press, 1974.

Hall, Edward. *The Union of the Two Noble and Illustrious Families of Lancaster and York.* London, 1548.

Harvey, William. *Anatomical Exercitations Concerning the Generation of Living Creatures.* London, 1653.

Harvey, William. *De motu cordis.* London, 1628.

Harvey, William. *The Anatomical Exercises "De Motu Cordis" and "De Circulatione Sanguinis" in English Translation.* Ed. Geoffrey Keynes. New York: Dover, 1995.

Heywood, John. *The Play of the Weather.* In *Tudor Interludes,* 141–80. Ed. Peter Happé. Harmondsworth, Middlesex: Penguin Books, 1972.

Holinshed, Raphael. *Holinshed's Chronicles: Richard II, 1398–1400, Henry IV and Henry V.* Ed. Alma Hansen and Robert Strachan Wallace. Oxford: Clarendon, 1923.

Holinshed, Raphael. *The First [Last] Volume of the Chronicles of England, Scotland, and Ireland.* London, 1577.

Holland, Henry. *The History of Adam, or the Four-Fold State of Man.* London, 1606.

Hooke, Robert. *Micrographia, or, Some Physiological Descriptions of Minute Bodies Made by Magnifying Glasses with Observations and Inquiries Thereupon.* London, 1665.

Hyperius, Andreas. *The Course of Christianity.* Trans. John Ludham. London, 1579.

Klaasen, Frank. *Making Magic in Elizabethan England: Two Early Modern Vernacular Books of Magic.* University Park: Pennsylvania State Press, 2019.

Lavater, Ludwig. *Of Ghosts and Spirits Walking By Night.* Trans. Robert Harrison. London, 1572.

Livy. *History of Rome*. Trans. B.O. Foster. 14 vols. Cambridge, MA: Harvard University Press, 1919–76.

Love, Nicolas. *The Mirror of the Blessed Life of Jesus Christ*. Ed. Michael G. Sargent. Exeter, UK: University of Exeter Press, 2004.

Lovejoy, Arthur Oncken, Gilbert Chinard, George Boas, and Ronald Salmon Crane, eds. *A Documentary History of Primitivism and Related Ideas*. Ann Arbor: Michigan University Press, 1935.

Luther, Martin. *A Very Comfortable and Necessary Sermon in These our Days*. London, 1570.

Luther, Martin. *Martin Luther's Basic Theological Writings*. Ed. Timothy F. Lull and William R. Russell. 3d. ed. Minneapolis: Fortress Press, 2012.

Luther, Martin. *The Last Will and Last Confession of Martin Luther's Faith Concerning the Principal Articles of Religion which are in Controversy*. London, 1543.

Luther, Martin. *Works*. Ed. Jaroslav Pelikan et al. 54 vols. St. Louis: Concordia Press, 1958–2024.

Lycosthenes, Conrad. *The Doom Warning All Men to Judgment*. Trans. and Ed. Stephen Batman. London, 1581.

Markham, Gervase. *The English Housewife*. Ed. Michael R. Best. Montreal: McGill-Queen's University, 1986.

Marlowe, Christopher. *Doctor Faustus*. Ed. David Scott Kastan. New York: W.W. Norton, 2005.

Merbeck, John. *A Book of Notes and Common Places, with their Expositions*. London, 1581.

Montemayor, George of. *A Critical Edition of Yong's Translation of George of Montemayor's "Diana" and Gil Polo's "Enamoured Diana."* Ed. Judith Kennedy. Oxford: Clarendon, 1968.

Moulton, Thomas. *The Mirror or Glass of Health*. London, 1580.

Munday, Anthony. *A Second and Third Blast of Retreat from Plays and Theaters*. London, 1580.

Nausea, Frederic. *All Blazing Stars in General*. Trans. Abraham Fleming. London, 1577.

Origen of Alexandria. *Patrologia cursus completes, series graeca*. Ed. J.-P. Minge. 166 vols. Paris, 1857–66.

Orosius, Paulus. *The Seven Books of History Against the Pagans*. Trans. Roy J. Deferrari. Washington, D.C.: Catholic University of America, 1964.

Ovid. *Metamorphoses*. Trans. Arthur Golding. Ed. Madeleine Forey. Baltimore: Johns Hopkins University Press, 2001.

Painter, William. *The Palace of Pleasure*. London, 1567.

Paracelsus. *Selected Writings*. Ed. Jolande Jacobi. Princeton: Princeton University Press, 1951.

Parmenides, *The Testimonia*. In *Early Greek Philosophy: Western Greek Thinkers*. 3 parts. Ed. and Trans. André Laks and Glenn W. Most. Cambridge, MA: Harvard University Press, 2016.

Peter of Limoges. *The Moral Treatise on the Eye*. Trans. Richard Newhauser. Toronto: Pontifical Institute of Mediaeval Studies, 2012.

Plato. *Collected Dialogues*. Ed. Edith Hamilton and Huntington Cairns. Princeton: Princeton University Press, 1961.

Platter, Felix. *The Structure and Function of the Human Body* (*De corporis humani structura et usu*). Basil, 1583.

Pliny the Elder. *A Summary of the Antiquities and Wonders of the World*. Trans. I.A. London, 1566.

Pliny the Elder. *The History of the World, Commonly Called the Natural History*. Trans. Philemond Holland. London, 1601.

Pliny the Elder. *Natural History*. Trans. H. Rackham. Cambridge University Press, 1938.

Plutarch. *Lives of the Noble Grecians and Romans*. Trans. Thomas North. London, 1579.

Pollard, Tanya, ed. *Shakespeare's Theater: A Sourcebook*. Oxford: Blackwell, 2004.

Primaudaye, Pierre de La. *The Second Part of the French Academy*. London, 1584.

Primaudaye, Pierre de La. *The Third Volume of the French Academy*. Trans. R. Dolman. London, 1601.

Pseudo-Dionysus. *The Complete Works*. Trans. Colm Luibhéid. Ed. Paul Rorem. New York: Paulist Press, 1987.

Puttenham, George. *The Art of English Poesy by George Puttenham: A Critical Edition*. Ed. Frank Whigham and Wayne A. Rebhorn. Ithaca: Cornell University Press, 2007.

Ralegh, Walter. *The History of the World*. London, 1614.

Ralegh, Walter. *The History of the World*. Ed. C.A. Patrides. Philadelphia: Temple University Press, 1971.

Rankins, William. *A Mirror for Monsters*. London, 1587.

Rastell, John. *Three Rastell Plays*. Ed. Richard Axton. Cambridge, MA: D.S. Brewer, 1979.

Reisch, Gregor. *Margarita philosophica*. In *Natural Philosophy Epitomised: A Translation of Books 8–11 of Gregor Reisch's "Philosophical Pearl" (1503)*. Ed. and Trans. Andrew Cunningham and Sachiko Kusukawa. Farnham, Surrey: Ashgate, 2010.

Scot, Reginald. *The Discovery of Witchcraft*. London, 1584.

Sennert, Daniel. *Thirteen Books of Natural Philosophy*. Trans. Nicholas Culpepper. London, 1660.

Seville, Isidore of. *The "Etymologies" of Isidore of Seville*. Ed. and Trans. Stephen A. Barney et al. Cambridge: Cambridge University Press, 2006.

Shakespeare, William. *Mr. William Shakespeare's Comedies, Histories, & Tragedies*. London, 1623.

Shakespeare, William. *Julius Caesar.* Ed. David Daniell. London: Bloomsbury, 1998.
Shakespeare, William. *King Lear.* Ed. Stanley Wells. Oxford: Oxford University Press, 2000.
Shakespeare, William. *Romeo and Juliet.* Ed. René Weis. London: Bloomsbury, 2012.
Shakespeare, William. *King Lear.* In *The Norton Shakespeare,* 2317–2570. 3d ed. Ed. Stephen Greenblatt, et. al. New York: Norton, 2016.
Shakespeare, William. *The New Oxford Shakespeare.* Ed. Gary Taylor, John Jowett, Terri Bourus, and Gabriel Egan. Oxford: Oxford University Press, 2016.
Shakespeare, William. *King Lear.* Ed. R.A. Foakes. London: Bloomsbury, 2018.
Sidney, Philip. *The Major Works.* Ed. Katherine Duncan-Jones. Oxford: Oxford University Press, 2002.
Statutes of the Realm, The. Ed. George Eyre and Andrew Strahan. 9 vols. London, 1810–1828.
Stubbes, Philip. *Anatomy of Abuses.* London, 1583
Suetonius. *Lives of the Caesars.* Trans. J.C. Rolfe. 2 vols. Cambridge, MA: Harvard University Press, 1914.
Tate, Nahum. *The History of King Lear: Acted at the Duke's Theatre.* London, 1681.
Turner, William. *A New Herbal.* London, 1551.
Tymme, Thomas. *Dialogue Philosophical.* London, 1612.
Tyndale, William, trans. *The New Testament, Diligently Corrected and Compared with the Greek.* Antwerp, 1534.
Tyndale, William. *The Obedience of a Christian Man.* London, 1528.
Vesalius, Andreas. *De humani corporis fabrica.* Basel, 1543.
Webster, John. *Academiarum examen, or, The Examination of Academies.* London, 1654.
Wickham, Glynn. *The Medieval Theater.* 3d. ed. Cambridge: Cambridge University Press, 1987.
Wright, Leonard. *A Display of Duty.* London, 1589
Zwingli, Ulrich. *Commentary on True and False Religion.* Zurich: Froschauer, 1525.

Secondary Sources

Ackerman, James S. "Alberti's Light." In *Studies in Late Medieval and Renaissance Painting in Honor of Millard Meiss,* 1:1-27. Ed. Irving Lavin and John Plummer. 2 vols. New York: NYU Press, 1977.
Ackerman, James S. *Origins, Imitation, Conventions: Representation in the Visual Arts.* Cambridge: MIT Press, 2002.
Adelman, Janet. *The Common Liar: An Essay on Antony and Cleopatra.* New Haven: Yale University Press, 1973.
Agamben, Giorgio. *Homo Sacer: Sovereign Power and Bare Life.* Trans. Daniel Heller-Roazen. Stanford: Stanford University Press, 1995.
Altman, Joel. *The Tudor Play of Mind: Rhetorical Inquiry and the Development of Elizabethan Drama.* Berkeley: University of California Press, 1978.

Arber, Agnes. *Herbals, Their Origin and Evolution: A Chapter in the History of Botany, 1470–1670*. Cambridge: Cambridge University Press, 1912.

Armbruster, Karla, and Kathleen R. Wallace, eds. *Beyond Nature Writing: Expanding the Boundaries of Ecocriticism*. Charlottesville, VA: University of Virginia Press, 2001.

Armstrong, Karen. *A History of God: The 4,000-year Quest of Judaism, Christianity, and Islam*. New York: Ballantine Books, 1994.

Armstrong, Philip. *Shakespeare's Visual Regime: Tragedy, Psychoanalysis and the Gaze*. London: Palgrave, 2000.

Arnheim, Rudolf. *Visual Thinking*. Berkeley: University of California Press, 1969.

Ashby, Richard. "'Pierced to the Soul': The Politics of the Gaze in *Richard II*." *Shakespeare* 11.2 (2015): 201–13.

Ashworth, Jr., William B. "Emblematic Natural History of the Renaissance." In *Cultures of Natural History*, 17–37. Ed. N. Jardine, J.A. Secord, and E.C. Spary. Cambridge: Cambridge University Press, 1996.

Astington, John H. *Stage and Picture in the English Renaissance: Mirror Up to Nature*. Cambridge: Cambridge University Press, 2017.

Auerbach, Eric. *Mimesis: The Representation of Reality in Western Literature*. Trans Willard R. Trask. Princeton: Princeton University Press, 1953.

Aulakh, Pavneet. "Seeing Things Through 'Images Sensible': Emblematic Similitudes and Sensuous Words in Francis Bacon's Natural Philosophy." *English Literary History* 81.4 (2014): 1149–72.

Baldwin, T.W. *William Shakespeare's Small Latin and Less Greek*. 2 vols. Urbana: University of Illinois Press, 1966.

Baldwin, T.W. "A Note upon William Shakespeare's Use of Pliny." In *Essays in Dramatic Criticism: The Parrott Presentation Volume*, 157–82. Ed. Hardin Craig. New York: Russell and Russell, 1967.

Barbour, Ian G. *Religion and Science: Historical and Contemporary Issues*. Revised ed. San Francisco: Harper, 1997.

Baskerville, Charles Read. "Conventional Features of Medwall's *Fulgens and Lucrece*." *Modern Philology* 24.4 (1927): 419–42.

Bate, Jonathan. *Shakespeare and Ovid*. Oxford: Clarendon, 1993.

Bate, Jonathan. "The Case for the Folio." www.rscshakespeare.co.uk. 2007.

Bate, Jonathan. *Soul of an Age: A Biography of the Mind of William Shakespeare*. New York: Random House, 2009.

Bates, Catherine. "Love and Courtship." *The Cambridge Companion to Shakespearean Comedy*. Ed. Alexander Leggatt. Cambridge: Cambridge University Press, 2002.

Battenhouse, Roy. *Shakespearean Tragedy: Its Art and Its Christian Premises*. Bloomington: Indiana University Press, 1969.

Bauer, Robert J. "Despite of Mine Own Nature: Edmund and the Orders, Cosmic and Moral." *Texas Studies in Literature and Language* 10.3 (1968): 359–66.

Beckwith, Sarah. *Signifying God: Social Relation and Symbolic Act in the York Corpus Christi Plays*. Chicago: University of Chicago Press, 2003.

Bellany, Alastair. *The Politics of Court Scandal in Early Modern England: News Culture and the Overbury Affair, 1603–1660*. Cambridge: Cambridge University Press, 2007.

Belsey, Catherine. *Shakespeare and the Loss of Eden: The Construction of Family Values in Early Modern Family Culture*. Houndsmills, UK: Macmillan, 1999.

Benjamin, Walter. "The Work of Art in the Age of Mechanical Reproduction. " In *Illuminations*, 217–51. Trans. Harry Zohn. Ed. Hannah Arendt. New York: Schocken Books, 1978.

Bennett, Jane. *Vibrant Matter: A Political Ecology of Things*. Durham, NC: Duke University Press, 2010.

Bevington, David. *From "Mankind" to Marlowe: Growth of Structure in The Popular Drama of Tudor England*. Cambridge, MA: Harvard University Press, 1962.

Bevington, David. *Tudor Drama and Politics*. Cambridge, MA: Harvard University Press, 1968.

Bevington, David, ed. *Medieval Drama*. Boston: Houghton Mifflin, 1975.

Blair, Ann. *Theater of Nature: Jean Bodin and Renaissance Science*. Princeton: Princeton University, 1997.

Blake, Liza. "Allegorical Causation and Aristotelian Physics in Henry Medwall's *Nature*." *Studies in English Literature* 55.2 (2015): 341–63.

Blank, Paula. *Shakespeare and the Mismeasure of Renaissance Man*. Ithaca: Cornell University Press, 2006.

Bloom, Allan. *Shakespeare's Politics*. Chicago: University of Chicago Press, 1964.

Bloom, Harold. *Shakespeare: The Invention of the Human*. New York: Riverhead, 1998.

Boehrer, Bruce. "The Privy and its Double: Scatology and Satire in Shakespeare's Theatre." In *A Companion to Shakespeare's Works, vol. 4: The Poems, Problem Comedies, Late Plays*, 69–88. Ed. Richard Dutton and Jean E. Howard. Oxford: Blackwell, 2003.

Boner, Patrick. *Kepler's Cosmological Synthesis: Astrology, Mechanism, and the Soul.* Leiden: Brill, 2013.

Bono, James J. *The Word of God and the Languages of Man: Interpreting Nature in Early Modern Science and Medicine: Vol. 1, Ficino to Descartes*. Madison: University of Wisconsin Press, 1995.

Boodin, J.E. "The Vision of Parmenides." *The Philosophical Review* 52.6 (1943): 578–589.

Bouwsma, William J. *John Calvin: A Sixteenth-Century Portrait*. Oxford: Oxford University Press, 1988.

Bowsher, Julian. *Rose Theatre: An Archaeological Discovery*. London: Museum of London, 1988.

Bowsher, Julian. *Shakespeare's London Theatreland: Archaeology, History and Drama*. London: Museum of London, 2012.

Bradley, A.C. *Shakespearean Tragedy: Lectures on "Hamlet," "Othello," "King Lear," "Macbeth."* 4th ed. London: Palgrave, 2007.

Bradley, Ritamary. "Backgrounds of the Title *Speculum* in Medieval Literature." *Speculum* 29.1 (1954): 100–15.

Brandt, Susan H. *Women Healers: Gender, Authority, and Medicine in Early Philadelphia*. Philadelphia: University of Pennsylvania Press, 2022.

Brayton, Dan. "Angling in the Lake of Darkness: Possession, Dispossession, and the Politics of Discovery in *King Lear*." *English Literary History* 70.2 (2003): 399–426.

Brenk, Frederick E. *In Mist Apparelled: Religious Themes in Plutarch's "Moralia" and "Lives."* Leiden: Brill, 1977.

Briggs, John C. *Francis Bacon and the Rhetoric of Nature*. Cambridge, MA: Harvard University Press, 1989.

Brokaw, Katherine Steele and Jay Zysk, eds. *Sacred and Secular Transactions in the Age of Shakespeare*. Evanston, IL: Northwestern University Press, 2019.

Brown, Deborah J. and Calvin G. Normore. *Descartes and the Ontology of Everyday Life*. Oxford: Oxford University Press, 2019.

Bryson, Norman, Michael Ann Holly, and Keith Moxey, eds. *Visual Culture: Images and Interpretations*. Hanover, N.H.: Wesleyan University Press, 1994.

Burnett, D. Graham. "Descartes and the Hyperbolic Quest: Lens Making Machines and Their Significance in the Seventeenth Century." *Transactions of the American Philosophical Society*, n.s. 95.3 (2005): 1–152.

Bushnell, Rebecca. "Vegetable Virtues." In *Shakespeare's Botanical Imagination*, 43–62. Ed. Susan C. Staub. Amsterdam: Amsterdam University Press, 2023.

Bynum, Caroline Walker. *Jesus as Mother: Studies in the Spirituality of the High Middle Ages*. Berkeley: University of California Press, 1982.

Bynum, William F. "The Great Chain of Being After Forty Years: An Appraisal." *The History of Science* 13.1 (1978): 1–28.

Calloway, Katherine. *Literature and Natural Theology in Early Modern England*. Cambridge: Cambridge University Press, 2023.

Camille, Michael. *Gothic Idol: Ideology and Image Making in Medieval Art*. Cambridge: Cambridge University Press, 1989.

Cantor, Paul. "Nature and Convention in *King Lear*." In *Poets, Princes, and Private Citizens: Literary Alternatives to Postmodern Politics*, 213–34. Ed. Joseph M. Knippenberg and Peter Augustine Lawler. New York: Rowman & Littlefield, 1996.

Carman, Charles H. *Leon Battista Alberti and Nicolas Cusanus: Towards an Epistemology of Vision for Italian Renaissance Art and Culture*. Farnham, Surrey: Ashgate, 2014.

Carroll, William C. "The Virgin Not: Language and Sexuality in Shakespeare." In *Shakespeare and Gender: A History,* 283–301. Ed. Ivo Kamps and Deborah E. Barker. London: Verso, 1995.

Carson, David. "The Dramatic Importance of Prodigies in *Julius Caesar.*" *English Language Notes* 2 (1965): 177–80.

Cavell, Stanley. "Epistemology and Tragedy: A Reading of *Othello.*" *Daedalus* 108.3 (1979): 27–43.

Cerasano, S.P. "Philip Henslowe, Simon Forman, and the Theatrical Community of the 1590s." *Shakespeare Quarterly* 44.2 (1993): 145–158.

Chalmers, Gordon K. "Hieroglyphs and Sir Thomas Browne." *Virginia Quarterly Review* 11.4 (1935): 547–60.

Cheney, Patrick. *Shakespeare, National Poet-Playwright.* Cambridge: Cambridge University Press, 2004.

Chiari, Sophie, and Mickaël Popelard, eds. *Spectacular Science, Technology, and Superstition in the Age of Shakespeare.* Edinburgh: Edinburgh University Press, 2017.

Christianson, John Robert. "Tycho Brahe's German Treatise on the Comet of 1577: A Study in Science and Politics." *Isis* 70.1 (1979): 132–40.

Clark, Stuart. *Vanities of the Eye: Vision in Early Modern European Culture.* Oxford: Oxford University Press, 2007.

Coakley, Sarah. *Re-thinking Dionysius the Areopagite.* Ed. Coakley and Charles M. Stang. Walton, MA: Wiley-Blackwell, 2009.

Colley, J.S. "*Fulgens and Lucrece*: Politics and Aesthetics." *Zeitschrift für Anglistik und Amerikanistik* 23.4 (1975): 322–30.

Collins, Francis S. *The Language of God: A Scientist Presents Evidence for Belief.* New York: Free Press, 2006.

Collinson, Patrick. *From Iconoclasm to Iconophobia: The Cultural Impact of the Second English Reformation.* Reading: University of Reading, 1986.

Cooper, Taryna. *Searching for Shakespeare.* New Haven: Yale University Press, 2006.

Copenhaver, Brian. "Magic." In *The Cambridge History of Science, vol. 3: Early Modern Science,* 518-40. Ed. Katharine Park and Lorraine Daston. Cambridge: Cambridge University Press, 2003.

Cosman, Brad. "*All's Well that Ends Well:* Shakespeare's Treatment of the Anal Fistula." *The Upstart Crow* 19 (1999): 78–95.

Crane, Mary Thomas. *Losing Touch with Nature: Literature and the New Science in 16th-Century England.* Baltimore: Johns Hopkins University Press, 2014.

Crawford, John W. "The Religious Question in *Julius Caesar.*" *Southern Quarterly* 15.3 (1977): 297–302.

Cummins, Juliet, and David Burchell, eds. *Science, Literature, and Rhetoric in Early Modern England.* Aldershot, Hampshire: Ashgate, 2007.

Cunningham, Andrew. "The Kinds of Anatomy." *Medical History* 19.1 (1975): 1–19.

Cunningham, Andrew. *The Anatomical Renaissance: The Resurrection of the Anatomical Projects of the Ancients.* Aldershot, UK: Scholar Press, 1997.

Curth, Louise Hill. *English Almanacs, Astrology, and Popular Medicine, 1550–1700.* Manchester: Manchester University Press, 2009.

Curtius, Ernst Robert. *European Literature and the Latin Middle Ages.* Trans. William R. Trask. London: Routledge and Kegan Paul, 1953.

Danby, John F. *Shakespeare's Doctrine of Nature: A Study of King Lear.* London: Faber and Faber, 1949.

Daston, Lorraine. "Baconian Facts, Academic Civility, and the Prehistory of Objectivity." *Annals of Scholarship* 8.3-4 (1991): 337–63.

Daston, Lorraine. "The Nature of Nature in Early Modern Europe." *Configurations* 6.2 (1998): 149–72.

Daston, Lorraine. "Observing Nature." In *Prints and the Pursuit of Knowledge in Early Modern Europe,* 125-61. Ed. Susan Dackerman. Cambridge: Harvard Art Museums, 2011.

Daston, Lorraine. *Against Nature.* Cambridge: MIT Press, 2019.

Daston, Lorraine, and Katharine Park. *Wonders and the Order of Nature, 1150–1750.* New York: Zone Books, 1998.

Daston, Lorraine, and Peter Galison. *Objectivity.* New York: Zone Books, 2007.

Davis, David J. *Seeing Faith, Printing Pictures: Religious Identity During the English Reformation.* Leiden: Brill, 2013.

Davis, Thomas J. "'The Truth of the Divine Words': Luther's Sermons on the Eucharist, 1521–28, and the Structure of Eucharistic Meaning." *The Sixteenth Century Journal* 30.2 (1999): 323–342.

Dawson, Anthony B. and Paul Yachnin. *The Culture of Playgoing in Shakespeare's England: A Collaborative Debate.* Cambridge: Cambridge University Press, 2001.

Dawson, Anthony. "The Secular Theater." In *Shakespeare and Religious Change,* 238–60. Ed. Kenneth J.E. Graham and Philip D. Collington. London: Palgrave, 2009.

Dear, Peter. *Revolutionizing the Sciences: European Knowledge in Transition, 1500–1700.* 3d ed. London: Springer Nature, 2019.

Debus, Allen. *The English Paracelsians.* Ann Arbor: University of Michigan Press, 1966.

Debus, Allen, ed. *Science, Medicine, and Society in the Renaissance.* New York: Science History Publications, 1972.

Debus, Allen G. *Man and Nature in the Renaissance.* Cambridge: Cambridge University Press, 1978.

Debus, Allen. *The Chemical Philosophy: Paracelsian Science and Medicine in the Sixteenth and Seventeenth Centuries.* New York: Dover, 2002.

Demaitre, Luke. *Medieval Medicine: The Art of Healing from Head to Toe.* London: Bloomsbury, 2013.

Derrida, Jacques. *Of Grammatology.* Trans. Gayatri Chakravorty Spivak. Baltimore: Johns Hopkins University Press, 1974.

Derrida, Jacques. *Dissemination.* Trans. Barbara Johnson. Chicago: University of Chicago Press, 1981.

Devereux, E.J. *A Bibliography of John Rastell.* Montreal: McGill University Press, 1999.

Diehl, Huston. "Horrid Image, Sorry Sight, Fatal Vision: The Visual Rhetoric of *Macbeth.*" *Shakespeare Studies* 16 (1983): 191–202.

Diehl, Huston. *Staging Reform, Reforming the Stage: Protestantism and Popular Theater in Early Modern England.* Ithaca: Cornell University Press, 1997.

Dillon, Matthew. *Omens and Oracles: Divination in Ancient Greece.* New York: Routledge, 2017.

Dobin, Howard. *Merlin's Disciples: Prophecy, Poetry, and Power in Renaissance England.* Stanford: Stanford University Press, 1990.

Dolan, T.P. "The Mass as Performance Text." In *From Page to Performance: Essays in Early English Drama*, 13–24. Ed. John A. Alford. East Lansing: Michigan State University Press, 1995.

Dollimore, Jonathan. *Radical Tragedy: Religion, Ideology, and Power in the Drama of Shakespeare and His Contemporaries.* Chicago: University of Chicago Press, 1984.

Doody, Aude. *Pliny's Encyclopedia: The Reception of the Natural History.* Cambridge: Cambridge University Press, 2010.

Drake, Stillman. "Galileo's Language: Mathematics and Poetry in a New Science. *Yale French Studies* 49.1 (1973): 13–27.

Drake, Stillman. *Essays on Galileo and the History and Philosophy of Science.* 3 vols. Toronto: University of Toronto Press, 1999.

Duncan-Jones, Katharine. *Shakespeare: An Ungentle Life.* London: Methuen Drama, 2010.

Dundas, Judith. "Shakespeare's Imagery: Emblem and the Imitation of Nature." *Shakespeare Studies* 16 (1983): 45–56.

Dyrness, William A. *Reformed Theology and Visual Culture: The Protestant Imagination from Calvin to Edwards.* Cambridge: Cambridge University Press, 2004.

Eagleton, Terry. *Sweet Violence: The Idea of the Tragic.* Malden, MA: Blackwell, 2003.

Eamon, William. *Science and the Secrets of Nature: Books of Secrets in Medieval and Early Modern Culture.* Princeton: Princeton University Press, 1996.

Easton, Stuart C. *Roger Bacon and His Search for a Universal Science: A Reconsideration of the Life and work of Roger Bacon in the Light of his Own Stated Purposes.* New York: Russell and Russell, 1971.

Economou, George. *The Goddess Natura in Medieval Literature.* Cambridge, MA: Harvard University Press, 1972.

Edwards, Kathryn A., ed. *Everyday Magic in Early Modern Europe*. London: Routledge, 2016.
Edwards, Philip. "Tragic Balance in *Hamlet*." *Shakespeare Survey* 36.1 (1983): 43–52.
Egan, Gabriel. "Hearing or Seeing a Play?: Evidence of Early Modern Theatrical Terminology." *Ben Jonson Journal* 8.2 (2001): 327–47.
Eggert, Katherine. "Nostalgia and the Not Yet Late Queen: Refusing Female Rule in Henry V." *English Literary History* 61.3 (1994): 523–50.
Eggert, Katherine. *Showing Like a Queen: Female Authority and Literary Experiment in Spenser, Shakespeare, and Milton*. Philadelphia: University of Pennsylvania Press, 2000.
Eggert, Katherine. *Disknowledge: Literature, Alchemy, and the End of Humanism in Renaissance England*. Philadelphia: University of Pennsylvania Press, 2016.
Eire, Carlos M.N. *War Against the Idols: The Reformation of Worship from Erasmus to Calvin*. Cambridge: Cambridge University Press, 1979.
Eisenstein, Elizabeth L. *The Printing Revolution in Early Modern Europe*. Cambridge: Cambridge University Press, 1983.
Elliott, Brent. "The World of the Renaissance Herbal." *Renaissance Studies* 25.1 (2011): 24–41.
Elmer, Peter. "Medicine, Religion, and the Puritan Revolution." In *The Medical Revolution of the Seventeenth Century*, 10–45. Ed. Roger French and Andrew Wear. Cambridge: Cambridge University Press, 1989.
Elmer, Peter, ed. *The Healing Arts: Health, Disease, and Society in Europe, 1500–1800*. Manchester: Manchester University Press, 2004.
Elton, William R. *"King Lear" and the Gods*. San Marino, CA: The Huntington Library, 1966.
Erickson, Peter and Clark Hulse, eds. *Early Modern Visual Culture: Representation, Race, and Empire in Renaissance England*. Philadelphia: University of Pennsylvania Press, 2000.
Evans, Malcolm. "Deconstructing Shakespeare's Comedies." In *Alternative Shakespeares*, 69-96. Ed. John Drakakis. London: Methuen, 1985.
Eversmann, Peter G.F. "What Did They See? Science and Religion in the Anatomical Theatres of the Sixteenth and Seventeenth Centuries." In *Quid est Secretum: Visual Representations of Secrets in Early Modern Europe, 1500–1700*, 260–84. Ed. Ralph Dekonicnk, Agnès Guiderdoni, and Walter S. Melion. Leiden: Brill, 2020.
Feerick, Jean. "Economies of Nature in Shakespeare." *Shakespeare Studies* 39 (2011): 32–42.
Field, Catherine. "'Sweet Practicer, thy Physic I will Try': Helena and her 'Good Receipt' in *All's Well that Ends Well*." In *All's Well that Ends Well: New Critical Essays*, 194–209. Ed. Gary Waller. New York: Routledge, 2007.
Findlen, Paula. "Jokes of Nature and Jokes of Knowledge: The Playfulness of Scientific Discourse in Early Modern Europe." *Renaissance Quarterly* 43.2 (1990): 292–331.

Findlen, Paula. "Anatomy Theaters, Botanical Gardens, and Natural History Collections." In *The Cambridge History of Science, Vol. 3: Early Modern Science*, 272–89. Ed. Katharine Park and Lorraine Daston. Cambridge: Cambridge University Press, 2006.

Findlen, Paula. "Natural History." In *The Cambridge History of Science, Vol. 3: Early Modern Science*, 435–68. Ed. Katharine Park and Lorraine Daston. Cambridge: Cambridge University Press, 2006.

Foucault, Michel. *The Order of Things: An Archaeology of the Human Sciences*. Trans. Alan Sheridan. New York: Vintage, 1973.

Franke, William. *On the Universality of What is Not: The Apophatic Turn in Critical Thinking*. Notre Dame: University of Notre Dame Press, 2020.

Franklin-Brown, Mary. *Reading the World: Encyclopedic Writing in the Scholastic Age*. Chicago: University of Chicago Press, 2012.

Freccero, John. "The Fig and the Laurel Tree: Petrarch's Poetics." *Diacritics* 5.1 (1975): 34–40.

Freedberg, David. *The Power of Images: Studies in the History and Theory of Response*. Chicago: University of Chicago Press, 1989.

Frye, Roland Mushat. *Shakespeare and Christian Doctrine*. Princeton: Princeton University Press, 1963.

Garber, Marjorie. *Dream in Shakespeare: From Metaphor to Metamorphosis*. New Haven: Yale University Press, 1974.

Garber, Marjorie. *Shakespeare's Ghost Writers: Literature as Uncanny Causality*. New York: Methuen, 1987.

Garber, Marjorie. "Dream and Interpretation: *Julius Caesar*." In *William Shakespeare's "Julius Caesar": Modern Critical Interpretations*, 43-52. Ed. Harold Bloom. Philadelphia: Chelsea House, 1988.

Gaukroger, Stephen. *The Emergence of a Scientific Culture: Science and the Shaping of Modernity, 1210–1685*. Oxford: Clarendon Press, 2006.

Gayk, Shannon. *Image, Text, and Religious Reform in Fifteenth-Century England*. Cambridge: Cambridge University Press, 2010.

Gee, Emma. *Ovid, Aratus, and Augustus: Astronomy in Ovid's "Fasti."* Cambridge: Cambridge University Press, 2000.

Gellrich, Jesse M. *The Idea of the Book in the Middle Ages: Language Theory, Mythology, and Fiction*. Ithaca: Cornell University Press, 1985.

Ghosh, Kantik. *The Wycliffite Heresy: Authority and the Interpretation of Texts*. Cambridge: Cambridge University Press, 2001.

Gil, Daniel Juan. "'Bare Life': Political Order and the Specter of Antisocial Being in Shakespeare's *Julius Caesar*." *Common Knowledge* 13.1 (2007): 67–69.

Gillen, Katharine A. "From Sacraments to Signs: The Challenges of Protestant Theatricality in John Bale's Biblical Plays." *Cahiers Élisabéthains* 80.1 (2011): 15–25.

Gillies, John. "The Question of Original Sin in *Hamlet*." *Shakespeare Quarterly* 64.4 (2013): 396-424.

Gilman, Ernest B. *The Curious Perspective: Literary and Pictorial Wit in the Seventeenth Century.* New Haven: Yale University Press, 1978.

Girard, René. "Lévi-Strauss, Frye, Derrida and Shakespearean Criticism." *Diacritics* 3.3 (1973): 34–48.

Girard, René. "Love Delights in Praises: A Reading of *The Two Gentlemen of Verona*." *Philosophy and Literature* 13.2 (1989): 231–47.

Glotfelty, Cheryll. *The Ecocriticism Reader: Landmarks in Literary Ecology.* Ed. Glotfelty and Harold Fromm. Athens: University of Georgia, 1996.

Goldberg, Jonathan. "Perspectives: Dover Cliff and the Conditions of Representation." *Poetics Today* 5.3 (1984): 537–48.

Goldberg, Jonathan. *Voice Terminal Echo: Postmodernism and English Renaissance Texts.* New York: Methuen, 1986.

Goldman, Michael. *Shakespeare and the Energies of Drama.* Princeton: Princeton University Press, 1972.

Goldstein, Leonard. *The Origin of Medieval Drama.* Madison: Fairleigh Dickinson University Press, 2004.

Gombrich, E.H. *The Story of Art.* 16th ed. Ann Arbor: University of Michigan Press, 1995.

Goodman, Katharine. "Inverting the Pietà in *King Lear*." In *Marian Moments in Early Modern British Drama*, 47–74. Ed. Regina Buccola and Lisa Hopkins. Farnham, UK: Ashgate, 2007.

Goulding, Robert. "Thomas Harriot's Optics, Between Experiment and Imagination: The Case of Mr. Bulkeley's Glass." *Archive for History of Exact Sciences* 68.2 (2014): 137–178.

Grady, Hugh. *Shakespeare, Machiavelli, and Montaigne: Power and Subjectivity from "Richard II" to "Hamlet."* Oxford: Oxford University Press, 2002.

Greenblatt, Stephen. "Invisible Bullets: Renaissance Authority and Its Subversion, *Henry IV* and *Henry V*." In *Political Shakespeare: New Essays in Cultural Materialism*, 18–47. Ed. Jonathan Dollimore and Alan Sinfied. Manchester: Manchester University Press, 1985.

Greenblatt, Stephen. "Shakespeare and the Exorcists." In *Shakespearean Negotiations: The Circulation of Social Energy in Renaissance England.* 94–128. Berkeley: University of California Press, 1988.

Groeneveld, Leanne. "'Lyke unto a Lyvelye Thyng': The Boxley Rood of Grace and Medieval Performance." In *Medieval Theatre Performance: Actors, Dancers, Automata, and their Audiences*, 197-214. Ed. Katie Normington and Philip Butterworth. Cambridge: D.S. Brewer, 2017.

Gudger, E.W. "Pliny's *Historia Naturalis:* The Most Popular Natural History Ever Published." *Isis* 6.3 (1924): 269–81.

Guérin, Sarah M. "Meaningful Spectacles: Gothic Ivories Staging the Divine." *The Art Bulletin* 95.1 (2013): 53–77.

Gurr, Andrew. *Shakespeare's Workplace: Essays on Shakespearean Theater.* Cambridge: Cambridge University Press, 2017.

Gwinn, John A. "The Letter Device in the First Act of *Two Gentlemen of Verona*." *Texas Studies in English* 20 (1940): 72–81.

Hacking, Ian. *Representing and Intervening: Introductory Topics in the Philosophy of Natural Science*. Cambridge: Cambridge University Press, 1983.

Håkansson, Håkan. "Tycho the Prophet: History, Astrology, and the Apocalypse in Early Modern Science." In *The Word and the World: Biblical Exegesis and Early Modern Science*, 137–56. Ed. Kevin Killeen and Peter J. Forshaw. New York: Palgrave Macmillan, 2007.

Halpern, Richard. *The Poetics of Primitive Accumulation: English Renaissance Culture and the Genealogy of Capital*. Ithaca: Cornell University Press, 1991.

Hamling, Tara. *Decorating the 'Godly' Household: Religious Art in Post-Reformation Britain*. New Haven: Yale University Press, 2010.

Hammill, Graham, and Julia Reinhard Lupton, eds. *Political Theology and Early Modernity*. Chicago: University of Chicago Press, 2012.

Hanning, R.W. "'You Have Begun a Parlous Pleye': The Nature and Limits of Dramatic Mimesis as a Theme in Four Middle English 'Fall of Lucifer' Cycle Plays." *Comparative Drama* 7.1 (1973): 22–50.

Hardison, O.B. *Christian Rite and Christian Drama in the Middle Ages: Essays in the Origin and Early History of Modern Drama*. Baltimore, MD: Johns Hopkins University Press, 1965.

Hardman, C.B. "'Our Drooping Country Now Erects Her Head': Nahum Tate's *History of King Lear*." *Modern Language Review* 95.4 (2000): 913–23.

Harkness, Deborah E. *The Jewel House: Elizabethan London and the Scientific Revolution*. New Haven: Yale University Press, 2007.

Harries, Karsten. *Infinity and Perspective*. Cambridge: MIT Press, 2002.

Harrison, Peter. "The Book of Nature and Early Modern Science." In *The Book of Nature in Early Modern and Modern History*, 1–26. Ed. Klaas van Berkel and Arjo Vanderjagt. Leuven: Peeters, 2006.

Harrison, Peter. *The Fall of Man and the Foundations of Science*. Cambridge: Cambridge University Press, 2001.

Harrison, Peter. *The Bible, Protestantism, and the Rise of Natural Science*. Cambridge: Cambridge University Press, 2007.

Harrison, Peter, ed. *The Cambridge Companion to Science and Religion*. Cambridge: Cambridge University Press, 2010.

Harvey, Elizabeth D. "The Touching Organ: Allegory, Anatomy, and the Renaissance Skin Envelope." In *Sensible Flesh: On Touch in Early Modern Culture*, 81–102. Ed. Harvey. Philadelphia: University of Pennsylvania Press, 2003.

Hayles, N. Katherine. *How We Became Posthuman: Virtual Bodies in Cybernetics, Literature, and Infomatics*. Chicago: University of Chicago Press, 1999.

Healy, Margaret. *Shakespeare, Alchemy, and the Creative Imagination: The Sonnets and "A Lover's Complaint*." Cambridge: Cambridge University Press, 2011.

Heilman, Robert Bechtold. *This Great Stage: Image and Structure in "King Lear."* Baton Rouge: Louisiana State University Press, 1948.

Hellegers, Desiree. *Handmaid to Divinity: Natural Philosophy, Poetry, and Gender in Seventeenth-Century England.* Norman: Oklahoma University Press, 2002.

Heller, Jack. "'Your Statue Spouting Blood': *Julius Caesar,* the Sacraments, and the Fountain of Life." In *Shakespeare's Christianity: The Protestant and Catholic Poetics of "Julius Caesar," "Macbeth," and "Hamlet,"* 77–93. Ed. Beatrice Batson. Waco, TX: Baylor University Press, 2006.

Hellman, Doris. "Bibliography of Tracts and Treatises on the Comet of 1577." *Isis* 22.1 (1934): 41–68.

Hellman, Doris. *The Comet of 1577 and Its Place in the History of Astronomy.* New York: AMS, 1971.

Heseler, Baldasar. *Andreas Vesalius' First Public Anatomy at Bologna, 1540: An Eyewitness Report.* Ed. and Trans. Ruben Eriksson. Uppsala and Stockholm: Almquist & Wiksells 1959.

Hillman, David. "Visceral Knowledge." In *The Body in Parts: Fantasies of Corporeality in Early Modern Europe,* 81–106. Ed. Hillman and Carla Mazzio. London: Routledge, 1997.

Hirsch-Reich, Beatrice, and Marjorie Reeves. *The Figurae of Joachim of Fiore.* Oxford: Oxford University Press, 1972.

Hirschfeld, Heather. "Hamlet's 'First Corse': Repetition, Trauma, and the Displacement of Redemptive Typology." *Shakespeare Quarterly* 54.4 (2003): 424–48.

Hoeniger, F. David. *Medicine and Shakespeare in the English Renaissance.* Newark: University of Delaware Press, 1992.

Hoff, Johannes. *The Analogical Turn: Rethinking Modernity with Nicholas of Cusa.* Grand Rapids, MI: Eerdmans, 2013.

Howell, Kenneth J. *God's Two Books: Copernican Cosmology and Biblical Interpretation in Early Modern Science.* Notre Dame, IN: University of Notre Dame Press, 2002.

Hughes, Ted. "An Idea Whose Time Has Come." *The Spectator* no. 7395 (21 March 1970): 378–9.

Hunt, Maurice. "*The Two Gentlemen of Verona* and the Paradox of Salvation." *Rocky Mountain Review of Language and Literature* 36.1 (1982): 5–22.

Hunt, Maurice. *Shakespeare's Religious Allusiveness: Its Play and Tolerance.* Aldershot, UK: Ashgate, 2004.

Hunt, Maurice. "Cobbling Souls in Shakespeare's *Julius Caesar.*" In *Shakespeare's Christianity: The Protestant and Catholic Poetics of "Julius Caesar," "Macbeth," and "Hamlet,"* 111–30. ed. Beatrice Batson. Waco, TX: Baylor University Press, 2006.

Hunter, G.K. *Dramatic Identities and Cultural Tradition: Studies in Shakespeare and His Contemporaries.* Ann Arbor: University of Michigan Press, 1978.

Hutton, Sarah. "Platonism in some Metaphysical Poets." In *Platonism and the English Imagination*, 163-77. Ed. Anna Baldwin and Sarah Hutton. Cambridge University Press, 1994.

Ilardi, Vincent. *Renaissance Vision from Spectacles to Telescopes.* Philadelphia: American Philosophical Association, 2007.

Irvine, Martin. *The Making of a Textual Culture: "Grammatica" and Literary Theory, 350–1100.* Cambridge: Cambridge University Press, 1994.

Iversen, Erik. *The Myth of Egypt and its Hieroglyphics in European Tradition.* Copenhagen: GED GAD, 1961.

Jackson, Ken and Arthur F. Marotti. "The Turn to Religion in Early Modern English Studies." *Criticism* 46.1 (2004): 167–190.

Jameson, Frederic. *The Antinomies of Realism.* London: Verso, 2015.

Jardine, Lisa. *Francis Bacon: Discovery and the Art of Discourse.* Cambridge: Cambridge University Press, 1974.

Jensen, Phebe. "Causes in Nature: Popular Astrology in *King Lear*." *Shakespeare Quarterly* 69.4 (2018): 205–27.

Johns, Adrian. *The Nature of the Book: Print and Knowledge in the Making.* Chicago: University of Chicago, 1998.

Jones, R.C. "The Stage World and the Real World in Medwall's *Fulgens and Lucrece*." *Modern Language Quarterly* 32.2 (1971): 131–42.

Kalas, Rayna. *Frame, Glass, Verse: The Technology of Poetic Invention in the English Renaissance.* Ithaca: Cornell University Press, 2007.

Kassell, Lauren. *Medicine and Magic in Elizabethan London: Simon Forman: Astrologer, Alchemist, and Physician.* Oxford: Clarendon, 2005.

Kastan, David Scott. "Proud Majesty Made a Subject: Shakespeare and the Spectacle of Rule." *Shakespeare Quarterly* 37.4 (1986): 459–475.

Kastan, David Scott. "'Holy Words' and 'Slypper Wit': John Bale's *King John* and the Poetics of Propaganda." In *Rethinking the Henrician Era*, 267-82. Ed. Peter C. Herman. Urbana and Chicago: University of Illinois Press, 1994.

Kauffmann, Thomas DaCosta. *The Mastery of Nature: Aspects of Art, Science, and Humanism in the Renaissance.* Princeton: Princeton University Press, 1993.

Kaula, David. "'Let Us Be Sacrificers': Religious Motifs in *Julius Caesar*." *Shakespeare Studies* 14 (1981): 197–214.

Keleher, Sarah. "Anatomy and Salvation History in Andreas Vesalius's *De humani corporis fabrica*." In *Rethinking Science and Religion in Early Modern Culture.* Ed. Jennifer Rust and Aaron Kitch. London: Bloomsbury, forthcoming.

Keller, Eve. "Producing Petty Gods: Margaret Cavendish's Critique of Experimental Science." *English Literary History* 64.2 (1997): 447–471.

Kemp, Martin. "Seeing and Picturing: Visual Representation in Twentieth-Century Science." In *Science in the Twentieth Century*, 361-90. Ed. John Krige and Dominique Pestre. Amsterdam: Hardwood Academic Publishers, 1997.

Kiefer, Frederick. "Love Letters in *The Two Gentlemen of Verona*." *Shakespeare Studies* 18 (1986): 65–85.

Kiefer, Frederick. *Shakespeare's Visual Theater: Staging Personified Characters.* Cambridge: Cambridge University Press, 2003.

Killeen, Kevin. *The Unknowable in Early Modern Thought: Natural Philosophy and the Poetics of the Ineffable.* Stanford: Stanford University Press, 2023.

King, Helen. *The Disease of Virgins: Green Sickness, Chlorosis and the Problems of Puberty.* London: Routledge, 2004.

Kinney, Arthur F. "Lear." *The Massachusetts Review* 17.4 (1976): 677–712.

Kitch, Aaron. "Medwall's 'Condycion': *Fulgens and Lucrece* and the New Tudor Drama." *Cahiers élisabéthains* 68 (2005): 1–8.

Klaassen, Frank. *Making Magic in Elizabethan England: Two Early Modern Vernacular Books of Magic.* University Park: Pennsylvania State University Press, 2019.

Klestinec, Cynthia. "Civility, Comportment, and Anatomy Theater: Girolamo Fabrici and His Medical Students in Renaissance Padua." *Renaissance Quarterly* 60.2 (2007): 434–63.

Klestinec, Cynthia. *Theaters of Anatomy: Students, Teachers, and Traditions of Dissection in Renaissance Venice.* Baltimore: Johns Hopkins University Press, 2011.

Knapp, Jeffrey. *Shakespeare's Tribe: Church, Nation, and Theater in Renaissance England.* Chicago: University of Chicago Press, 2002.

Knight, G. Wilson. *Shakespeare and Religion: Essays of Forty Years.* 1949; Reprt. London: Taylor & Francis, 2013.

Knight, Nicholas W. "*Julius Caesar*: A Case of Pre- and Post-Christian Story." *Christianity and Literature* 28.4 (1979): 27–35.

Kobialka, Michal. *This Is My Body: Representational Practices in the Early Middle Ages.* Ann Arbor: University of Michigan Press, 1999.

Kocher, Paul. "Lady Macbeth and the Doctor." *Shakespeare Quarterly* 5.4 (1954): 341–9.

Kocher, Paul. *Science and Religion in Elizabethan England.* New York: Octagon Books, 1969.

Koerner, Joseph Leo. *The Reformation of the Image.* London: Reaktion, 2004.

Korhonen, Anu. "Washing the Ethiope White: Conceptualizing Black Skin in the English Renaissance." In *Black Africans in Renaissance Europe*, 94–112. Ed. T.F. Earle and K.J.P. Lowe. Cambridge: Cambridge University Press, 2005.

Kott, Jan. *Shakespeare our Contemporary.* Trans. Boleslaw Taborski. London: Methuen, 1964.

Koyré, Alexandre. "Galileo and Plato." *Journal of the History of Ideas* 4.4 (1943): 400–28.

Kusukawa, Sachiko. *Picturing the Book of Nature: Image, Text, and Argument in Sixteenth-Century Human Anatomy and Medical Botany.* Chicago: University of Chicago Press, 2012.

Lacan, Jacques. "The Mirror Stage." In *Écrits: A Selection,* 1–7. Trans. Alan Sheridan. New York: Norton, 1977.

Lafont, Robert. *Anthropologie de l'écriture*. Paris: Centre Georges Pompidou, 1984.

Lake, Peter. *Hamlet's Choice*. New Haven: Yale University Press, 2020.

Lamp, Kathleen S. *A City of Marble: The Rhetoric of Augustan Rome*. Columbia: University of South Carolina Press, 2013.

Latour, Bruno. *The Politics of Nature: How to Bring the Sciences into Democracy*. Cambridge: Harvard University Press, 2004.

LeVert, Laurelle. "'Crucifye hem, Crucifye hem': The Subject and Affective Response in Middle English Passion Narratives." *Essays in Medieval Studies* 14 (1998): 73–90.

Levin, Harry. *The Question of Hamlet*. New York: Viking, 1961.

Lindberg, David C. "Science and the Early Church." In *God and Nature: Historical Essays on the Encounter Between Christianity and Science*, 19–48. Ed. Lindberg and Ronald L. Numbers. Berkeley: University of California Press, 1986.

Lindberg, David C. "Medieval Science and Its Religious Context." *Osiris* 10 (1995): 60–79.

Lindberg, David C. "Science and the Medieval Church." In *The Cambridge History of Science, vol. 2: Medieval Science*, 268–85. Ed. Lindberg and Michael H. Shank. Cambridge: Cambridge University Press, 2013.

Lindeman, Mary. *Medicine and Society in Early Modern Europe*. Cambridge: Cambridge University Press, 2010.

Linderski, Jerzy. "The Augural Law." *Aufstieg und Niedergang der römischen Welt: Geschicte und Kultur Roms im Spiegel der neueren Forschung* 2.16.3 (1975): 2146–2312.

Linderski, Jerzy. "Cicero and Roman Divination." *La Parola del Passato* 37 (1982): 12–38.

Long, Pamela O. "Objects of Art / Objects of Nature: Visual Representation and the Investigation of Nature." In *Merchants and Marvels: Commerce, Science, and Art in Early Modern Europe*, 63-82. Ed. Pamela H. Smith and Paula Findlen. New York: Routledge, 2002.

Lovejoy, Arthur O. *The Great Chain of Being: A Study of the History of an Idea*. Cambridge: Harvard University Press, 1936.

Lovejoy, Arthur O. *Essays in the History of Ideas*. 1948; reprt. Baltimore: Johns Hopkins University Press, 2019.

Lovejoy, Arthur O., Gilbert Chinard, George Boas, and Ronald S. Crane, eds. *Documentary History of Primitivism and Related Ideas*. Baltimore: Johns Hopkins University Press, 1935.

Lupton, Julia. *Afterlives of the Saints: Hagiography, Typology, and Renaissance Literature*. Stanford: Stanford University Press, 1996.

Lüthy, Christoph. "What Does a Diagram Prove that Other Images Do Not? Images and Imagination in the Kepler-Fludd Controversey." In *Image, Imagination, and Cognition: Medieval and Early Modern Theory and Practice*, 227-74. Ed. Lüthy et. al. Leiden: Brill, 2018.

Lynch, Michael, and Steve Woolgar, eds. *Representation in Scientific Practice*. Cambridge, Mass: MIT Press, 1990.
MacDonald, Ronald R. "Playing Till Doomsday: Interpreting *Antony and Cleopatra*." *English Literary Renaissance* 15.1 (1985): 78–99.
Mack, Maynard. *Shakespeare in Our Time*. Berkeley: University of California Press, 1965.
Maclean, Ian. "Foucault's Renaissance Episteme Reassessed: An Aristotelian Counterblast." *Journal of the History of Ideas* 59.1 (1998): 149–66.
Malay, Jessica L. *Prophecy and Sibylline Imagery in the Renaissance*. New York: Routledge, 2010.
Malcolmson, Cristina. *Studies of Skin Color in the Early Royal Society: Boyle, Cavendish, Swift*. Farnham, Surrey: Ashgate, 2013.
Mandelbrote, Scot. "Early Modern Biblical Interpretation and the Emergence of Science." *Science & Christian Belief* 23.2 (2011): 99–113.
Marchitello, Howard. *The Machine in the Text: Science and Literature in the Age of Shakespeare and Galileo*. Oxford University Press, 2011.
Marion, Jean-Luc. *The Crossing of the Visible*. Stanford: Stanford University Press, 2004.
Marrapodi, Michele, ed. *Shakespeare and the Visual Arts: The Italian Influence*. London: Routledge, 2017.
Martin, Graham and Stephen Regan. "*King Lear*." In *Shakespeare: Texts and Contexts*, 241–76. Ed. Kiernan Ryan. New Yok: St. Martin's, 2000.
Massai, Sonia. "Nahum Tate's Revision of Shakespeare's *King Lears*." *Studies in English Literature* 40.3 (2000): 435–50.
Masten, Jeffrey. "*Two Gentlemen of Verona*." In *A Companion to Shakespeare's Works, Vol. 3: The Comedies*, 266–88. Ed. Jean Howard and Richard Dutton. Malden, MA: Wiley, 2003.
Matthews, Steven. *Theology and Science in the Thought of Francis Bacon*. Aldershot, Hampshire: Ashgate, 2008.
Mayer, Jean-Christoph. *Shakespeare's Hybrid Faith: History, Religion, and the Stage*. New York: Palgrave Macmillan, 2006.
Mazzio, Carla. "Shakespeare and Science, c. 1600." *South Central Review* 2.1 & 2 (2009): 1–23.
McCray, W. Patrick. *Glassmaking in Renaissance Venice: The Fragile Craft*. London: Taylor & Francis, 2017.
McDermott, Joseph Peter, and Peter Burke, eds. *The Book Worlds of East Asia and Europe, 1450–1850: Connections and Comparisons*. Hong Kong: Hong Kong University Press, 2015.
McDonald, Mark Allen. *Shakespeare's "King Lear" with "The Tempest": The Discovery of Nature and the Shakespearean Recovery of Classical Natural Right*. Dallas: University Press of America, 2004.
McGinn, Bernard. "Hildegard of Bingen as Visionary and Exegete." In *Hildegard von Bingen in ihrem historischen Umfeld*, 321–50. Ed. Alfred Haverkamp. Mainz: Philipp von Zabern, 2000.

McGinn, Bernard. "Theologians as Trinitarian Iconographers." In *The Mind's Eye: Art and Theological Argument in the Middle Ages*, 186–207. Ed. Jeffrey F. Hamburger and Anne-Marie Bouché. Princeton: Princeton University Press, 2005.

McLuhan, Marshall. *The Gutenberg Galaxy: The Making of Typographic Man*. Toronto: University of Toronto Press, 2011.

McMillin, Scott. "Shakespeare's *Richard II*: Eyes of Sorrow, Eyes of Desire." *Shakespeare Quarterly* 35.1 (1984): 40–52.

Meer, Jitse van der, and Richard J. Oosterhoof. "God, Scripture, and the Rise of Modern Science (1200–1700): Notes in the Margin of Harrison's Hypothesis." In *Nature and Scripture in the Abrahamic Religions: Up to 1700*. 2 vols. Ed. Jitse van der Meer and Scott Mandelbrote. Leiden: Brill, 2008. 2:363–96.

Merchant, Carolyn. *The Death of Nature: Women, Ecology and the Scientific Revolution*. San Francisco: Harper & Row, 1980.

Miles, Margaret. "Theology, Anthropology, and the Human Body in Calvin's *Institutes of the Christian Religion*." *The Harvard Theological Review* 74.3 (1981): 303–23.

Milner, Matthew. *The Senses and the English Reformation*. Farnham, Surrey: Ashgate, 2011.

Milward, Peter. *Biblical Influences in Shakespeare's Great Tragedies*. Bloomington: Indiana University Press, 1987.

Mitchell, Fiona. *Monsters in Greek Literature: Aberrant Bodies in Ancient Greek Cosmogony, Ethnography, and Biology*. London: Taylor & Francis, 2021.

Mitchell, W.J.T. *Iconology: Image, Text, Ideology*. Chicago: University of Chicago Press, 1986.

Mitchell, W.J.T. *Picture Theory: Essays in Verbal and Visual Representation*. Chicago: University of Chicago Press, 1994.

Mitchell, W.J.T. "Interdisciplinarity and Visual Culture." *Art Bulletin* 77.4 (1995): 540–44.

Montagnes, Bernard. *The Doctrine of the Analogy of Being According to Thomas Aquinas*. Trans. E. M. Macierowski, P. Vandevelde, and Andrew Tallon. Milwaukee: Marquette University Press, 2004.

Moore, James W. *Capitalism in the Web of Life: Ecology and the Accumulation of Capital*. London: Verso, 2015.

Moran, Bruce. *Paracelsus: An Alchemical Life*. Chicago: Reaktion Books, 2017.

Morgan, Edward. *The Incarnation of the Word: The Theology of Language of Augustine of Hippo*. London: Continuum, 2010.

Morton, Timothy. *Ecology Without Nature: Rethinking Environmental Aesthetics*. Cambridge, MA: Harvard University Press, 2009.

Moynihan, Robert D. "Stars, Portents, and Order in *Julius Caesar*." *Modern Language Studies* 7.2 (1977): 26–31.

Mulryne, J.R. and Margaret Shewring, eds. *Shakespeare's Globe Rebuilt*. Cambridge University Press, 1997.

Nauert, Charles G., Jr. "Humanists, Scientists, and Pliny: Changing Approaches to a Classical Author." *The American Historical Review* 84.1 (1979): 72-85.
Ndiaye, Noémie. *Scripts of Blackness: Early Modern Performance Culture and the Making of Race*. Philadelphia: University of Pennsylvania Press, 2022.
Nelson, Robert S. ed. *Visuality Before and Beyond the Renaissance*. Cambridge: Cambridge University Press, 2000.
Newman, Barbara. *Medieval Crossover: Reading the Secular Against the Sacred*. South Bend, IN: University of Notre Dame Press, 2013.
Newman, William. *Promethean Ambitions: Alchemy and the Quest to Perfect Nature*. Chicago: University of Chicago Press, 2004.
Nicholson, Max. *Environmental Revolution*. London: Hodder & Stoughton, 1970.
North, John. "Some Weak Links in the Great Chain of Being." In *Empirical Logic and Public Debate: Essays in Honor of Else M. Barth*, 105–20. Ed. Erik C.W. Krabbe, Renée José Dalitz, and Pier A. Smit. Leiden: Brill, 2023.
Numbers, Ronald, ed. *Galileo Goes to Jail and Other Myths about Science and Religion*. Cambridge: Harvard University Press, 2009.
Nunn, Hillary M. *Staging Anatomies: Dissection and Spectacle in Early Stuart Tragedy*. Aldershot, UK: Ashgate, 2005.
Nuttall, A.D. *Shakespeare the Thinker*. New Haven: Yale University Press, 2007.
O'Connell, Michael. *The Idolatrous Eye: Iconoclasm & Theater in Early Modern England*. Oxford: Oxford University Press, 2000.
Ogilvie, Brian W. "The Many Books of Nature: Renaissance Naturalists and Information Overload." *Journal of the History of Ideas* 64.1 (2003): 29–40.
Olson, Roberta J.M. "...And They Saw Stars: Renaissance Representations of Comets and Pretelescopic Astronomy." *Art Journal* 44.3 (1984): 216–224.
Ong, Walter J. *Ramus, Method, and the Decay of Dialogue: From the Art of Discourse to the Art of Reason*. Cambridge: Harvard University Press, 1983.
Orgel, Stephen. "Shakespeare Imagines a Theater." *Poetics Today* 5.3 (1984): 549–61.
Orgel, Stephen. *Impersonations: The Performance of Gender in Shakespeare's England*. Cambridge: Cambridge University Press, 1996.
Orlemanski, Julie. *Symptomatic Subjects: Bodies, Medicine, and Causation in the Literature of Late Medieval England*. Philadelphia: University of Pennsylvania Press, 2019.
Orne, Nicholas. *Education and Society in Medieval and Renaissance England*. Ann Arbor: University of Michigan Press, 1989.
Orrell, John. *The Quest for Shakespeare's Globe*. Cambridge. Cambridge University Press, 1983.
Ortiz, Jared. *Deification in the Latin Patristic Tradition*. Ed. Ortiz. Washington, D.C.: Catholic University of America Press, 2019.
Osborne, Thomas. "Faith, Philosophy, and the Nominalist Background to Luther's Defense of the Real Presence." *Journal of the History of Ideas* 63.1 (2002): 63–82.

Osler, Margaret. "The Gender of Nature and the Nature of Gender in Early Modern Natural Philosophy." In *Men, Women, and the Birthing of Modern Science*, 71–85. Ed. Judith P. Zinsser. Dekalb: Northern Illinois University Press, 2005.

Pagel, Walter. *Paracelsus: An Introduction to Philosophical Medicine in the Era of the Renaissance*. 2d. ed. New York: Karger, 1982.

Park, Katharine. "Dissecting the Female Body: From Women's Secrets to the Secret of Nature." In *Crossing Boundaries: Attending to Early Modern Women*, 29–47. Ed. Adele F. Seeff and Jane Donawerth. Newark: University of Delaware Press, 2001.

Park, Katharine. *Secrets of Women: Gender, Generation, and the Origins of Human Dissection*. London: Zone Books, 2006.

Parker, John. *The Aesthetics of Antichrist from Christian Drama to Christopher Marlowe*. Ithaca: Cornell University Press, 2007.

Parshall, Karen Hunger, Michael T. Walton, and Bruce T. Moran, eds. *Bridging Traditions: Alchemy, Chemistry, and Paracelsian Practices in the Early Modern Era*. Kirksville, MO: Truman State University Press, 2015.

Pauwels, Luc, ed. *Visual Cultures of Science: Rethinking Representational Practices in Knowledge Building and Science Communication*. Hanover, NH: Dartmouth College Press, 2006.

Peterson, Kaara. "Elizabeth I's Virginity and the Body of Evidence: Jonson's Notorious Crux." *Renaissance Quarterly* 68.3 (2015): 840–871.

Peterson, Kaara. "The Ring's the Thing: Elizabeth I's Virgin Knot and *All's Well That Ends Well*." *Studies in Philology* 113.1 (2016): 101–31.

Phillips, John. *The Reformation of Images: Destruction of Art in England, 1535–1660*. Berkeley: University of California Press, 1973.

Phoon, Adrian. "'A Vision Fair and Fortunate': Ideology, Politics, and Selfhood in *Julius Caesar*." *Sydney Studies in English* 30.1 (2004): 21–41.

Poole, William. "Francis Lodwick's Creation: Theology and Natural Philosophy in the Early Royal Society." *Journal of the History of Ideas* 66.2 (2005): 245–63.

Principe, Lawrence. *The Aspiring Adept: Robert Boyle and His Alchemical Quest*. Princeton: Princeton University Press, 1998.

Principe, Lawrence. *The Scientific Revolution: A Very Short Introduction*. Oxford: Oxford University Press, 2011.

Principe, Lawrence. *The Secrets of Alchemy*. Chicago: University of Chicago Press, 2013.

Prior, Moody. *The Drama of Power: Studies in Shakespeare's History Plays*. Evanston, Ill: Northwestern University Press, 1973.

Pye, Christoper. "The Betrayal of the Gaze: Theatricality and Power in Shakespeare's *Richard II*." *English Literary History* 55.3 (1988): 575–98.

Pye, Christoper. *The Regal Phantasm: Shakespeare and the Politics of Spectacle*. London: Routledge, 1990.

Rabb, Felix. *The English Face of Machiavelli: A Changing Interpretation 1500–1700.* London: Routledge and Keegan Paul, 1964.

Raber, Karen. *Shakespeare and Posthumanist Theory.* London: Arden Shakespeare, 2020.

Rabkin, Norman. *Shakespeare and the Problem of Meaning.* Chicago: University of Chicago Press, 1981.

Rackin, Phyllis. *The Stages of History: Shakespeare's English Chronicles.* Ithaca: Cornell University Press, 1990.

Reisner, Noam. *Milton and the Ineffable.* Oxford: Oxford University Press, 2009.

Reiss, Timothy. *Tragedy and Truth.* New Haven: Yale University Press, 1980.

Reiss, Timothy. *The Discourse of Modernism.* Ithaca: Cornell University Press, 1982.

Remien, Peter. *The Concept of Nature in Early Modern English Literature.* Cambridge: Cambridge University Press, 2019.

Ribner, Irving. "Bolingbroke, a True Machiavellian." *Modern Language Quarterly* 9.2 (1948): 177–84.

Rice, Julian. "*Julius Caesar* and the Judgment of the Senses," *Studies in English Literature* 13.2 (1973): 238–55.

Riggs, David. *The World of Christopher Marlowe.* New York: Henry Holt, 2014.

Ritchey, Sara. *Holy Matter: Changing Perceptions of the Material World in Late Medieval Christianity.* Ithaca: Cornell University Press, 2014.

Robertson, Kellie. *Nature Speaks: Medieval Literature and Aristotelian Philosophy.* Philadelphia: University of Pennsylvania Press, 2017.

Roe, John. "'Character' in Plutarch and Shakespeare: Brutus, Julius Caesar, and Mark Antony." In *Shakespeare and the Classics*, 173–87. Ed. Charles Martindale and A.B. Taylor. Cambridge: Cambridge University Press, 2004.

Rorty, Richard. *Philosophy and the Mirror of Nature.* Princeton: Princeton University Press, 1979.

Rose, Mark. "Conjuring Caesar: Ceremony, History, and Authority in 1599." *English Literary Renaissance* 19.3 (1989): 291–304.

Roychoudhury, Suparna. "Melancholy, Ecstasy, Phantasma: The Pathologies of *Macbeth*." *Modern Philology* 111.2 (2013): 205–30.

Rudwick, Martin J.S. "The Emergence of a Visual Language for Geological Science, 1760–1840." *History of Science* 14.3 (1976): 148–95.

Russell, Bertrand. *Our Knowledge of the External World as a Field in Modern Philosophy.* London: Routledge, 2009.

Sabatier, Armelle. *Shakespeare and Visual Culture: A Dictionary.* London: Bloomsbury, 2017.

Sawday, Jonathan. *The Body Emblazoned: Dissection and the Human Body in Renaissance Culture.* London: Routledge, 1995.

Sawyer, Robert. *Marlowe and Shakespeare: The Critical Rivalry.* London: Palgrave, 2017.

Schanzer, Ernest. "The Tragedy of Shakespeare's Brutus." *English Literary History* 22.1 (1955): 1–15.

Schildgen, Brenda Deen. *Divine Providence, A History: The Bible, Virgil, Orosius, Augustine, and Dante*. London: Continuum, 2012.

Schmitt, Charles B. "Experience and Experiment: A Comparison of Zarabella's view with Galileo's in *De Motu*." *Studies in the Renaissance* 16 (1969): 80–138.

Schreiner, Susan. *The Theater of His Glory: Nature and the Natural Order in the Thought of Calvin*. Grand Rapids, MI: Baker Academic, 1991.

Schutz, Andrea. "Absent and Present Images: Mirrors and Mirroring in John Gower's *Confessio Amantis*." *Chaucer Review: A Journal of Medieval Studies and Literary Criticism* 34.1 (1999): 107–24.

Scott, Charlotte. *Shakespeare's Nature: From Cultivation to Culture*. Oxford: Oxford University Press, 2014.

Scott, William O. "Proteus in Spenser and Shakespeare: The Lover's Identity." *Shakespeare Studies* 1 (1965): 283–93.

Serjeanstson, Richard. "Francis Bacon and the 'Interpretation of Nature' in the Late Renaissance." *Isis* 105.4 (2014): 681–705.

Shannon, Laurie. "Poor, Bare, Forked: Animal Sovereignty, Human Negative Exceptionalism, and the Natural History of *King Lear*." *Shakespeare Quarterly* 60.2 (2009): 168–96.

Shapin, Steven, and Simon Schaffer. *Leviathan and the Air Pump: Hobbes, Boyle, and the Experimental Life*. 1985; Rev. ed. Princeton: Princeton University Press, 2017.

Shippey, Tom. "The Protean Allegory of Matter in *The Two Gentlemen of Verona*." *Allegorica: A Journal of Medieval and Renaissance Literature* 27.1 (2011): 20–30.

Shuger, Deborah. *Habits of Thought in the English Renaissance*. Los Angeles: University of California Press, 1990.

Shuger, Deborah. *Political Theologies in Shakespeare's England: The Sacred and the State in "Measure for Measure."* London: Palgrave Macmillan, 2001.

Siegfried, Brandie R., and Lisa T. Sarasohn, eds. *God and Nature in the Thought of Margaret Cavendish*. Aldershot, Hampshire: Ashgate, 2014.

Simpson, James. *Reform and Cultural Revolution, Vol. 2: 1350–1547*. Oxford: Oxford University Press, 2002.

Sinfield, Alan. "Hamlet's Special Providence." *Shakespeare Survey* 33 (1980): 89–98.

Singer, Thomas C. "Sir Thomas Browne's 'Emphaticall decussation, or fundamentall figure': Geometrical Hieroglyphs and *The Garden of Cyrus*." *English Literary Renaissance* 17.1 (1987): 86–90.

Siraisi, Nancy G. *Medieval and Early Renaissance Medicine: An Introduction to Knowledge and Practice*. Chicago: University of Chicago Press, 1990.

Siraisi, Nancy G. "Medicine, 1450–1620, and the History of Science." *Isis* 103.3 (2012): 491–514.
Sisson, C.J. "Shakespeare's Helena and Dr. William Harvey." *Essays and Studies* 13 (1960): 1–20.
Smith, A. Mark. *From Sight to Light: The Passage from Ancient to Modern Optics.* Chicago: University of Chicago Press, 2015.
Smith, Jeffrey Chipps. *Nuremberg, A Renaissance City, 1500–1618.* Austin: University of Texas Press, 1983.
Smith, Pamela. *The Body of the Artisan: Art and Experience in the Scientific Revolution.* Chicago: University of Chicago Press, 2004.
Smith, Pamela. "Art, Science, and Visual Culture in Early Modern Europe." *Isis* 97.1 (2006): 83–100.
Snow, C.P. *The Two Cultures and the Scientific Revolution.* Cambridge: Cambridge University Press, 1988.
Soellner, Rolf. "Shakespeare, Aristotle, Plato, and the Soul." *Shakespeare-Jarbuch* 104 (1968): 56–71.
Soellner, Rolf. *Shakespeare's Patterns of Self-knowledge.* Columbus: Ohio State University Press, 1972.
Sohmer, Stephen T. *Shakespeare's Mystery Play: The Opening of the Globe Theater, 1599.* Manchester University Press, 1999.
Solberg, Emma Maggie. "Mirror Images: Mary and Hermione, Idolatry and Iconoclasm." In *Sacred and Secular Transactions in the Age of Shakespeare,* 221-40. Ed. Katherine Steele Brokaw and Jay Zysk, 2019.
Southern, Richard. *The Staging of Plays before Shakespeare.* London: Faber and Faber, 1973.
Speaight, Robert. *Nature in Shakespearean Tragedy.* New York: Collier, 1962.
Spiller, Elizabeth. *Science, Reading, and Renaissance Literature: The Art of Making Knowledge, 1580–1670.* Cambridge: Cambridge University Press, 2004.
Spinks, Jennifer. *Monstrous Births and Visual Culture in Sixteenth-Century Germany.* Farnham, Surrey: Ashgate, 2009.
Spinrad, Phoebe S. "The Fall of the Sparrow and the Map of Hamlet's Mind." *Modern Philology* 102.4 (2005): 453–477.
Staub, Susan C, ed. *Shakespeare's Botanical Imagination.* Amsterdam: Amsterdam University Press, 2023.
Steffen, William H. *Anthropocene Theater and the Shakespearean Stage.* Oxford: Oxford University Press, 2023.
Stephenson, Henry Thew. *Shakespeare's London.* New York: Henry Holt, 1905.
Stone, Lawrence. "Literacy and Education in England, 1640–1900." *Past and Present* 42.1 (1969): 69–139.
Strocchia, Sharon T. *Forgotten Healers: Women and the Pursuit of Health in Late Renaissance Italy.* Cambridge: Harvard University Press, 2019.
Struck, Peter T. *Divination and Human Nature: A Cognitive History of Intuition in Classical Antiquity.* Princeton: Princeton University Press, 2016.

Sutton, Clive. "'Nullius in Verba' and 'Nihil in Verbis': Public Understanding of the Role of Language in Science." *The British Journal for the History of Science* 27.1 (1994): 55–64.

Swarbrick, Steven. *The Environmental Unconscious: Ecological Poetics from Spenser to Milton*. Minneapolis: University of Minneapolis Press, 2023.

Swinburne, Richard. "The Social Theory of the Trinity." *Religious Studies* 54.3 (2018): 419–37.

Tallis, Raymond. *The Enduring Significance of Parmenides: Unthinkable Thought*. New York: Continuum, 2007.

Tassi, Marguerite A. *The Scandal of Images: Iconoclasm, Eroticism, and Painting in Early Modern English Drama*. Cranbury, NJ: Associated University Press, 2005.

Taylor, Charles. *A Secular Age*. Cambridge, MA: Belknap, 2007.

Taylor, Dennis. *Shakespeare and the Elizabethan Reformation: Literary Negotiation of Religious Difference*. Lanham, MD: Lexington Books, 2022.

Taylor, Gary. "Monopolies, Show Trials, Disaster, and Invasion: *King Lear* and Censorship." In *The Division of the Kingdoms*, 75–117. Ed. Taylor and Michael Warren. Oxford: Oxford University Press, 1983.

Thompson, Peter. *Shakespeare's Theater*. 2d ed. New York: Routledge, 1992.

Thompson, Rosemarie Garland. *Extraordinary Bodies: Figuring Physical Disability in American Culture and Literature*. New York: Columbia University Press, 1996.

Traister, Barbara Howard. *The Notorious Astrological Physician of London: Works and Days of Simon Forman*. Chicago: University of Chicago Press, 2010.

Tribble, Evelyn. "Sight and Spectacle." In *Shakespeare's Theaters and the Effects of Performance*, 237–52. Ed. Farah Karim-Cooper and Tiffany Stern. London: Arden, 2013.

Tribble, Evelyn. "Pretty and Apt: Boy Actors, Skill, and Embodiment." In *The Oxford Handbook of Shakespeare and Embodiment: Gender, Sexuality, and Race*. Ed. Valerie Traub. Oxford: Oxford University Press, 2016.

Truax, Elizabeth. "Emblematic Pictures for the Less Privileged in Shakespeare's England." In *Emblem, Iconography, and Drama*, 147–67. Ed. Clifford Davidson, Luis R. Gámez, and John H. Stroupe. Kalamazoo, Mich.: Medieval Institute Publications, 1995.

Turner, Henry S. "*King Lear* Without: The Heath," *Renaissance Drama* 28 (1997): 161–93.

Turner, Henry S. *The English Renaissance Stage: Geometry, Poetics, and the Practical Spatial Arts 1580–1630*. Oxford: Oxford University Press, 2006.

Turner, Henry S., ed. *Early Modern Theatricality*. Oxford: Oxford University Press, 2013.

Vanagt, Katrien. "Early Modern Medical Thinking on Vision and the Camera Obscura: V.F. Plempius's *Ophthalmographia*." In *Blood, Sweat and Tears: The Changing Concepts of Physiology from Antiquity into Early Modern Europe*, 569–94. Ed. Claus Zittel, H. F. J. Horstmanshoff, and Helen King. Leiden: Brill, 2012.

Vance, Eugene. "Saint Augustine: Language as Temporality." In *Mimesis: From Mirror to Method, Augustine to Descartes*, 20–35. Ed. and Intro. John D. Lyons and Stephen G. Nichols, Jr. Hanover, NH: Dartmouth University Press, 1982.

Velz, John. "Undular Structure in *Julius Caesar*." *Modern Language Review* 66.1 (1971): 21–30.

Vickers, Brian. *Francis Bacon and Renaissance Prose*. Cambridge: Cambridge University Press, 1968.

Waldron, Jennifer. *Reformations of the Body: Idolatry, Sacrifice, and Early Modern Theater*. New York: Palgrave, 2013.

Walker, Julia M., ed. *Dissing Elizabeth: Negative Representations of Gloriana*. Durham, NC: Duke University Press, 1998.

Warburg, Aby. "Pagan-Antique Prophecy in Words and Images." In *The Renewal of Pagan Antiquity: Contributions to the Cultural History of the European Renaissance*, 597–697. Los Angeles: Getty Research Institute, 1999.

Warner, John Harley. "The History of Science and the Sciences of Medicine." *Osiris*. n.s. 10 (1995): 164–193.

Watson, Robert N. *The Rest is Silence: Death as Annihilation in the English Renaissance*. Berkeley: University of California Press, 1994.

Watson, Robert N. *Back to Nature: The Green and the Real in the Late Renaissance*. Philadelphia: University of Pennsylvania Press, 2006.

Watt, Tessa. *Cheap Print and Popular Piety, 1550–1640*. Cambridge: Cambridge University Press, 1991.

Wedel, Theodore Otto. *The Mediæval Attitude Toward Astrology*. New Haven: Yale University Press, 1920.

Weeks, Andrew. *Paracelsus: Speculative Theory and the Crisis of the Early Reformation*. Albany: SUNY Press, 1997.

Weeks, Sophie. "Francis Bacon and the Art-Nature Distinction." *Ambix* 54.2 (2007): 101–29.

Weimann, Robert. "Laughing With the Audience: *The Two Gentlemen of Verona* and the Popular Tradition of Comedy." *Shakespeare Survey* 22 (1969): 35–42.

Weimann, Robert. *Shakespeare and the Popular Tradition: Studies in the Social Dimension of Dramatic Form and Function*. Ed. Robert Schwartz. Baltimore: Johns Hopkins University, 1987.

Weimann, Robert. "Representation and Performance: The Uses of Authority in Shakespeare's Theater," *PMLA* 107.3 (1992): 497–510.

Weinstock, Stefan. *Divus Julius*. Oxford: Clarendon, 1971.

Weis, René. *Shakespeare Revealed: A Biography*. Edinburgh: John Murray, 2007.

Wells, Stanley. *Shakespeare & Co.: Christopher Marlowe, Thomas Dekker, Ben Jonson, Thomas Middleton, John Fletcher and the Other Players in His Story*. New York: Pantheon, 2006.

West, William. *Theatres and Encyclopedias in Early Modern Europe*. Cambridge: Cambridge University Press, 2002.

Westman, Robert S. "Nature, Art, and Psyche: Jung, Pauli, and the Kepler-Fludd Polemic." In *Occult and Scientific Mentalities in the Renaissance*, 177–230. Ed. Brian Vickers. Cambridge: Cambridge University Press, 1986.

Whaley, Leigh. *Women and the Practice of Medical Care in Early Modern Europe, 1400–1800*. New York: Palgrave, 2011.

Wiesner, Merry. "Luther and Women: The Death of the Two Marys." In *Feminist Theology: A Reader*, 123–37. Ed. Ann Loades. London: SPCK, 1990.

Willes, Margaret. *A Shakespeare Botanical*. Oxford: Bodleian Library, 2015.

Williams, Linda. *Hard Core: Power, Pleasure, and the 'Frenzy of the Visible.'* Berkeley: University of California Press, 1999.

Williams, Mark. *Fiery Shapes: Celestial Portents and Astrology in Ireland and Wales, 700–1700*. Oxford University Press, 2010.

Wilson, Catherine. *The Invisible World: Early Modern Philosophy and the Invention of the Microscope*. Princeton: Princeton University Press, 1995.

Wilson, Richard. *Secret Shakespeare: Studies in Theatre, Religion, and Resistance*. Manchester: Manchester University Press, 2004.

Wilson, Richard. *Christopher Marlowe*. London: Taylor & Francis, 2014.

Witmore, Michael. *Shakespearean Metaphysics*. London: Continuum, 2008.

Wittkower, Rudolf. *Allegory and the Migration of Symbols*. Boulder, CO: Westview, 1977.

Wragge-Morley, Alexander. *Aesthetic Science: Representing Nature in the Royal Society of London, 1650–1720*. Chicago: University of Chicago Press, 2020.

Wright, L.B. "Notes on *Fulgens and Lucrece*: New Light on the Interlude." *Modern Language Notes* 41.2 (1926): 97–100.

Young, Karl. "The *Speculum majus* of Vincent of Beauvais." *The Yale University Library Gazette* 5.1 (1930): 5–13.

Young, Karl. *Drama of the Medieval Church*. 2 vols. Oxford: Clarendon Press, 1962.

Young, R.V. "Hope and Despair in *King Lear*." In *"King Lear": New Critical Essays*, 253–77. Ed. Jeffrey Kahan. New York: Routledge, 2008.

Ziebart, Meredith. "Cusanus and Nominalism." In *Nicholas of Cusa and Times of Transition: Essays in Honor of Gerald Christianson*, 219–39. Ed. Donald Duclow, Jason Aleksander, and Thomas M. Izbicki. Leiden: Brill, 2018.

Index

A. Kitch, *Shakespeare's Theater of Nature*, Palgrave Studies in Literature, Science and Medicine,
https://doi.org/10.1007/978-3-031-78082-0

J

K

L

M

R

T

V

The manufacturer's authorised representative in the EU is Springer Nature Customer Service Centre GmbH, Europaplatz 3, 69115 Heidelberg, Germany. If you have any concerns regarding our products, please contact ProductSafety@springernature.com

Printed and bound by CPI Group (UK) Ltd, Croydon, CR0 4YY

01/10/2026

02225954-0007